Lecture Notes in Mathematics

A collection of informal reports and seminars
Edited by A. Dold, Heidelberg and B. Eckmann, Zürich

313

Konrad Jörgens
Fakultät für Mathematik der Universität München, München/BRD

Joachim Weidmann
Fachbereich Mathematik der Universität Frankfurt, Frankfurt a. M./ BRD

Spectral Properties of Hamiltonian Operators

Springer-Verlag
Berlin · Heidelberg · New York 1973

AMS Subject Classifications (1970): 35J10, 35P05, 81-02, 81A09, 81A10, 81A81

ISBN 3-540-06151-7 Springer-Verlag Berlin · Heidelberg · New York
ISBN 0-387-06151-7 Springer-Verlag New York · Heidelberg · Berlin

Offsetdruck: Julius Beltz, Hemsbach/Bergstr.

Contents

Contents

Introduction

In this paper we are going to study Hamiltonian operators of quantum mechanical systems composed of N arbitrary particles. These operators are defined in $L_2(R^{3N})$ and are usually assumed to be of the form

$$(I) \qquad S = \sum_{k=1}^{N} S_k + \sum_{h<k} W_{hk} ,$$

where

$$S_k u(x) = \frac{1}{2\mu_k} \sum_{j=1}^{3} \left(i\frac{\partial}{\partial x_{kj}} + e_k b_j(x_k)\right)^2 u(x) + e_k v(x_k) u(x),$$

$$W_{hk} u(x) = w_{hk}(x_h - x_k) u(x),$$

$x_k = (x_{k1}, x_{k2}, x_{k3})$ the R^3-coordinate of the k-th particle, $x = (x_1, \dots, x_N) \in R^{3N}$; μ_k is the mass and e_k the charge of the k-th particle; $b = (b_1, b_2, b_3)$ is the vector potential of a magnetic field, v the electric potential and w_{hk} the interaction potential of the h-th and the k-th particle.

If the functions b_j and v all vanish, a change of variables leads to a separation of the center of mass motion and allows the study of the so called "internal Hamiltonian", an operator which acts only on the "internal" (relative) coordinates of the system; for the definitions of these notions see sections 5 and 6. We are interested in both, the total Hamiltonian (if there is a non-trivial external field) and the internal Hamiltonian (in case of a free system).

Actually we shall study a much more general class of operators

$$\text{(II)} \qquad S = T + \sum_{n=1}^{N} \sum_{j_1<\ldots<j_n} V_{j_1\ldots j_n} + \sum_{n=2}^{N} \sum_{j_1<\ldots<j_n} W_{j_1\ldots j_n},$$

where

$$T = - \sum_{k=1}^{N} \frac{1}{2\mu_k} \Delta_k$$

is the operator of kinetic energy of all particles (Δ_k the Laplacean with respect to $x_k = (x_{k1}, x_{k2}, x_{k3})$) and for every n-tuple $j_1\ldots j_n$ with $1 \leq j_1 <\ldots< j_n \leq N$, $V_{j_1\ldots j_n}$ and $W_{j_1\ldots j_n}$ represent the interaction of the particles $j_1,\ldots,j_n$ with an external field and with each other respectively. The conditions on the operators $W_{j_1\ldots j_n}$ are such, that, in case all $V_{j_1\ldots j_n}$ vanish, the separation of the center of mass motion is possible; in this case we shall consider the internal Hamiltonian.

Let us now first give a short account of the development of the theory of N-particle Hamiltonians.

The first rigorous definition of the self-adjoint realization of the Hamiltonian (total and internal) for any system composed of a finite number of particles interacting with each other through potential energies (e.g. Coulomb potentials) was given by Kato [16]. More general results in this direction were given later on by Stummel [25] and Ikebe-Kato [10]; some of these results are stated in section 2. For Hamiltonians of systems in regions with boundary see Jörgens [11].

Once the Hamiltonian has been defined one would like to know the spectral properties of these operators. It is generally expected that these operators are bounded from below and that the spectrum consists of a half line $[\mu,\infty)$ ($\mu \leq 0$) and at most countably many eigenvalues below μ, clustering at most at the point μ; for all physically relevant problems this seems to be true. In these cases the half line $[\mu,\infty)$ is the essential spectrum (see section 1).

Mainly three problems have been studied in the literature:

(1) Determination of the essential spectrum of Hamiltonians.

(2) Number of discrete eigenvalues, i.e. eigenvalues not belonging to the essential spectrum (by number we mean: zero, finitely many or infinitely many).

(3) Non-existence of eigenvalues contained in the essential spectrum or especially the non-existence of positive eigenvalues.

To the problems (2) and (3) there are solutions only for rather special cases. To problem (2): Kato [17], Zislin and Sigalov [37-42], Simon [23] showed that for many interesting systems there are infinitely many discrete eigenvalues; see also Uchiyama [26-30]. To problem (3): Kato [18], Weidmann [31, 32] and Albeverio [1] proved the non-existence of positive eigenvalues for some special cases.

In contrast to problems (2) and (3) problem (1) can be solved for very general systems. In this paper we are concerned only with this question.

In 1960 Zislin [38] studied Hamiltonians of type (I), where $b_j = 0$ and v, w_{hk} are Coulomb like potentials, $v \leq 0$, $w_{hk} \geq 0$; for example the Hamiltonians of atoms with fixed (infinitely heavy) nucleus are of this type. Zislin's result was: The essential spectrum $\sigma_e(S)$ is equal to $[\mu, \infty)$ where

$$\mu = \min \{\mu_j : j = 1, \ldots, N\},$$

$$\mu_j = \min \{\lambda \in \sigma(S^{(j)})\}$$

and $S^{(j)}$ is the Hamiltonian of the system consisting of all but the j-th particle. The proof of this result has been simplified by Jörgens in his unpublished report [12]; at the same time he generalized the result to systems with $b_j \neq 0$, where v, w_{hk} and b_j tend to zero near infinity in some generalized sense only (e.g. our corollary 3.14). In 1965 Zislin and Sigalov [41, 42] proved the same result for the restrictions of the Hamiltonian to the symmetry subspaces corresponding to irreducible representations of the symmetry group of the system (the symmetry group consisting of permutations of the identical particles and rotation-reflection symmetry). All these and some other results are collected in a survey article by Sigalov [22].

For arbitrary N-particle systems (without a condition on the sign of the interactions) only the internal Hamiltonian has been studied so far. In 1959 Zislin [37] announced a result for the internal Hamiltonian of an N-particle system with Coulomb interactions. In connection with scattering problems in 1964-65 van Winter [36] studied such systems with interaction potentials decreasing faster than Coulomb potentials. In 1966 Hunziker [9] considered N-particle systems where the interactions are locally square integrable tending to zero near infinity. In all these cases the result is: the essential spectrum of the internal Hamiltonian is $[\mu, \infty)$, where

$$\mu = \min \{\mu(Z_1) + \mu(Z_2)\}$$

and the min is taken over all decompositions of the system into two non-trivial subsystems Z_1 and Z_2, and $\mu(Z_i)$ is the greatest lower bound of the (total or internal) Hamiltonian of the system Z_i. Van Winter's and Hunziker's proofs essentially coincide: both use a certain functional equation due to Weinberg [33] (see also Hunziker [8]) and van Winter [36]. Van Winter proves that the operator involved in this equation is of Hilbert-Schmidt-Type, while Hunziker proves the compactness of this operator, and only this is needed. This compactness is essentially a consequence of the fact that the interactions are compact relative to the Laplacean with respect to the coordinates which are involved (for a precise definition see our section 6). This fact has been used by Combes [4]; he proved the same result for arbitrary n-particle interactions which are relatively compact in this sense. Van Winter's result has been generalized to systems with certain spin-orbit interactions by Brascamp and van Winter [3].

Later Zislin [39, 40] generalized his results to the restrictions of the internal Hamiltonian of an N-particle system to the subspaces corresponding to the irreducible representations of the symmetry group (permutation of identical particles and rotation reflection symmetry). Again the essential spectrum was found to be an interval $[\mu, \infty)$ and μ is given by means of sums of greatest lower bounds of Hamiltonians restricted to certain symmetry subspaces. The exact statement is rather difficult and we refer the reader to our general statement in section 11.

Zislin's proof is (as the proof of his earlier results) based on a decomposition of the coordinate space $R^{3(N-1)}$. Recently Balslev [2] gave a new proof of these results based on the Weinberg-van Winter equation; i.e. he extended Hunziker's proof to problems involving symmetries. Balslev claims that his proof also works in the more general situation of arbitrary relatively compact interactions; on the other hand in his introduction he postpones the consideration of nonlocal interactions to a later publication.

Similar results for a quite different situation have been given by Simon [24]; he considered N-particle systems with so called Rollnik-potentials, which are allowed to have local singularities such that the Hamiltonians can be defined only by means of quadratic forms and Friedrichs extension. Using a modified Weinberg-van Winter equation he proves the same results (also considering some symmetries) for this case. Finally let us mention Kato's survey article [19] which contains almost all the results mentioned above, and Jörgens' lecture notes [14].

Our aim is to give characterizations of the essential spectra of internal and total Hamiltonians of N-particle systems under very general conditions; we also consider the restrictions of these operators to symmetry subspaces. Our results contain all earlier results besides the result given by Simon; this is actually not comparable with our result.

Let us now outline the contents of this paper. In section 1 we give Weyl's characterization of the essential spectrum of self-adjoint operators by means of singular sequences; we extend these results to essentially self-adjoint operators, which is very useful for applications to differential operators. These results have already been used in [12] without proof.

Section 2 gives an introduction to the theory of Schrödinger operators. The material of this section is also taken from [12] and is probably well known to all specialists. For completeness we give all proofs.

In section 3 we study symmetric perturbations V of essentially self-adjoint operators T in $L_2(R^m)$. Here we introduce the important notion of T-smallness at infinity for perturbations V and we prove several interesting properties of these perturbations. This notion is the basis for the rest of this paper. All relevent results seem to be new. Examples are given in section 4; we show that all perturbations of the Laplacean, which have been considered earlier, are contained in our class. The pseudo-differential operator of theorem 4.13 even shows that our class is much larger.

In section 5 we introduce operators acting only on part of the variables. This section is mainly of technical nature and is a preparation for section 6, where we give our general definition of N-particle Hamiltonians. The symmetry group of an N-particle Hamiltonian is studied in section 7; in contrast to Zislin [39, 40] and Balslev [2] we do not assume that the full rotation-reflection group is contained in the symmetry group but only some compact subgroup. We give also several results on the structure of the corresponding symmetry subspaces which we use later on.

In section 8 we study the spectrum of the Hamiltonians of free systems restricted to symmetry subspaces. In this connection we use and prove some general results which may be of interest independent of our work (e.g. theorem 8.11).

An estimate of the essential spectrum of operators of the form $S=T+V$ ($T=-\Delta$) is proved in section 9. This is the essential part of one half of our main results. In section 10 we determine the essential spectra of N-particle Hamiltonians restricted to symmetry subspaces for systems with an external field (total Hamiltonian). The analogous result for the internal Hamiltonian is stated in section 11; the proof is given in section 12. In the appendix we indicate how our results can be extended to N-particle systems with spin, i.e. to similar operators in $(L_2(\mathbb{R}^{3N}))^k$ or $(L_2(\mathbb{R}^{3(N-1)}))^k$ for some $k \in \mathbb{N}$.

Similar to Zislin-Sigalov [38-42] and Jörgens [12] we use a decomposition of the coordinate space. Hunziker's very elegant proof [9] by means of the Weinberg-van Winter equation cannot be applied because our interactions are not relatively compact.

1. Spectra and essential spectra of selfadjoint and essentially selfadjoint operators

In this section we define the spectrum of an essentially selfadjoint operator A in a Hilbert space as the spectrum of its selfadjoint extension $\overline{A}$, and we characterize certain subsets of the spectrum by directly referring to A rather than to $\overline{A}$. This approach has advantages when dealing with differential operators. Since it seems that there exists no comprehensive account of this topic in the literature, we give complete proofs.

For a selfadjoint operator A in a Hilbert space H we denote by D(A) and R(A) the *domain* and the *range* of A respectively. The *resolvent set* $\rho(A)$ is the set of all $\lambda \in \mathbb{C}$ such that $\lambda - A$ has a bounded inverse $(\lambda - A)^{-1}$ with domain H; an equivalent definition is $\rho(A) = \{\lambda \in \mathbb{C} : R(\lambda - A) = H\}$. The *spectrum* $\sigma(A) = \mathbb{C} \setminus \rho(A)$ is a closed subset of the real line $\mathbb{R}$. Every isolated point of $\sigma(A)$ is an eigenvalue of A. The *discrete spectrum* of A is the set $\sigma_d(A)$ of all isolated eigenvalues of finite multiplicity. Its complement $\sigma_e(A) = \sigma(A) \setminus \sigma_d(A)$ is called the *essential spectrum*; it consists of all limit points of $\sigma(A)$ and of all isolated points which are eigenvalues of infinite multiplicity. Consequently $\sigma_e(A)$ is a closed subset of $\mathbb{R}$, whereas $\sigma_d(A)$ need not be closed.

For the rest of this section let A denote an *essentially selfadjoint* operator in the Hilbert space H (i.e. A is symmetric and the closure $\overline{A} = A^{**}$ is selfadjoint; in this case $\overline{A}$ is the unique selfadjoint extension of A). Let $E(\cdot)$ denote the spectral family of $\overline{A}$.

<u>1.1 Theorem.</u> The following statements are equivalent:

(1) $\lambda \in \sigma(\overline{A})$,

(2) $E(\lambda+\varepsilon) - E(\lambda-\varepsilon) \neq 0$ for all $\varepsilon > 0$,

(3) there exists a sequence (u_n) in $D(\overline{A})$ such that $\liminf \|u_n\| > 0$ and $(\lambda-\overline{A})u_n \to 0$,

(3') there exists a sequence (u_n) in $D(A)$ such that $\liminf \|u_n\| > 0$ and $(\lambda-A)u_n \to 0$.

<u>Proof.</u> (1) $\Rightarrow$ (2): Assume (1) and $E(\lambda+\varepsilon) - E(\lambda-\varepsilon) = 0$ for some $\varepsilon > 0$. Then $\lambda-\overline{A}$ has a bounded inverse defined by

$$(\lambda-\overline{A})^{-1}f = \int_{-\infty}^{\infty} (\lambda-t)^{-1}dE(t)f$$

for all $f \in H$, hence $\lambda \in \rho(\overline{A})$, a contradiction.

(2) $\Rightarrow$ (3): Choose $u_n \in (E(\lambda+\frac{1}{n}) - E(\lambda-\frac{1}{n}))H$ with $\|u_n\|=1$; it follows $u_n \in D(\overline{A})$ and $\|(\lambda-\overline{A})u_n\| \leq \frac{1}{n}$, hence $\liminf \|u_n\|=1$ and $(\lambda-\overline{A})u_n \to 0$.

(3) $\Rightarrow$ (3'): Take a sequence (u_n) according to (3). Since $\overline{A}$ is the closure of A, for every n there is a $v_n \in D(A)$ such that

$$\|u_n - v_n\| + \|\overline{A}u_n - Av_n\| \leq \frac{1}{n}.$$

It follows that

$$\|v_n\| \geq \|u_n\| - \frac{1}{n} \quad \text{and} \quad \|(\lambda-A)v_n\| \leq \|(\lambda-\overline{A})u_n\| + (1+|\lambda|)\frac{1}{n}$$

hence $\liminf \|v_n\| > 0$ and $(\lambda-A)v_n \to 0$.

(3') $\Rightarrow$ (1): Assume (3') and $\lambda \in \rho(\overline{A})$. Choose a sequence (u_n) according to (3'). Then since $\lambda-\overline{A}$ has a bounded inverse, $(\lambda-\overline{A})u_n = (\lambda-A)u_n \to 0$ implies $u_n \to 0$, a contradiction to $\liminf \|u_n\| > 0$.

<u>1.2 Theorem.</u> The following statements are equivalent:

(1) $\lambda \in \sigma_e(\overline{A})$,

(2) $\dim (E(\lambda+\varepsilon) - E(\lambda-\varepsilon))H = \infty$ for all $\varepsilon > 0$,

(3) there exists a sequence (u_n) in $D(\overline{A})$ such that $\langle u_n | u_m \rangle = \delta_{nm}$ and $(\lambda-\overline{A})u_n \to 0$,

(3') the same as (3) but with $\overline{A}$ replaced by A,

(4) there exists a sequence (u_n) in $D(\overline{A})$ such that $\liminf \|u_n\| > 0$, $u_n \rightharpoonup 0$ and $(\lambda-\overline{A})u_n \to 0$,

(4') the same as (4) but with $\overline{A}$ replaced by A.

Proof. (1) ⇒ (2): if $\lambda \in \sigma_e(\overline{A})$ then either λ is an (isolated) eigenvalue of infinite multiplicity or λ is a limit point of $\sigma(\overline{A})$. In the first case we have $\dim\,(E(\lambda+0) - E(\lambda-0))H = \infty$ and consequently $\dim\,(E(\lambda+\varepsilon) - E(\lambda-\varepsilon))H = \infty$ for all $\varepsilon > 0$. In the second case choose a sequence (λ_n) of mutually different points $\lambda_n \in \sigma(\overline{A})$ converging to λ, then choose positive numbers ε_n such that the intervals $[\lambda_n-\varepsilon_n, \lambda_n+\varepsilon_n]$ are mutually disjoint and consequently $\varepsilon_n \to 0$. Now for every $\varepsilon > 0$ there is a $n(\varepsilon)$ such that $[\lambda_n-\varepsilon_n, \lambda_n+\varepsilon_n] \subset]\lambda-\varepsilon, \lambda+\varepsilon[$ for all $n \geq n(\varepsilon)$, and hence

$$\dim\,(E(\lambda+\varepsilon)-E(\lambda-\varepsilon))H \geq \sum_{n=n(\varepsilon)}^{\infty} \dim(E(\lambda_n+\varepsilon_n)-E(\lambda_n-\varepsilon_n))H = \infty .$$

(2) ⇒ (3): Since $\dim\,(E(\lambda+\frac{1}{n}) - E(\lambda-\frac{1}{n}))H = \infty$ for all n, by induction we can find for all n an element $u_n \in (E(\lambda+\frac{1}{n}) - E(\lambda-\frac{1}{n}))H$ such that $\|u_n\| = 1$ and $\langle u_n | u_m \rangle = 0$ for $m < n$. It follows $u_n \in D(\overline{A})$, $\langle u_n | u_m \rangle = \delta_{nm}$ and $\|(\lambda-\overline{A})u_n\| \leq \frac{1}{n}$ for all n and m, hence $(\lambda-\overline{A})u_n \to 0$.

(3) ⇒ (4'): Take a sequence (u_n) according to (3). Since $\overline{A}$ is the closure of A, for every n there is a $v_n \in D(A)$ such that

$$\|u_n-v_n\| + \|\overline{A}u_n-Av_n\| \leq \frac{1}{n}.$$

It follows that

$$\|v_n\| \geq 1-\frac{1}{n} \quad \text{and} \quad \|(\lambda-A)v_n\| \leq \|(\lambda-\overline{A})u_n\| + (1+|\lambda|)\frac{1}{n},$$

hence $\liminf \|v_n\| > 0$ and $(\lambda-A)v_n \to 0$. Finally $u_n \rightharpoonup 0$ because (u_n) is an orthonormal sequence, and $u_n - v_n \to 0$ by construction, hence $v_n \rightharpoonup 0$.

(4') ⇒ (3'): Choose a sequence (u_n) according to (4') and assume (without restriction of generality) $\|u_n\| \geq 2$ for all n. We are going to select a subsequence (u_{n_k}) by induction as follows: If $u_{n_1},\dots,u_{n_k}$ are chosen denote by D_k the subspace spanned by these elements and by P_k the orthogonal projection onto D_k. Since $u_n \rightharpoonup 0$ and P_k is compact, we have $P_k u_n \to 0$ for $n \to \infty$. Therefore we can find $n_{k+1} > n_k$ such that

$$\|P_k u_{n_{k+1}}\| \sup \{1+\|(\lambda-A)u\| : u \in D_k,\ \|u\|=1\} \le \frac{1}{k}.$$

In particular we have $\|P_k u_{n_{k+1}}\| \le 1$ so that

$$v_{k+1} = \|u_{n_{k+1}} - P_k u_{n_{k+1}}\|^{-1}(u_{n_{k+1}} - P_k u_{n_{k+1}})$$

is defined, has norm 1 and is orthogonal to D_k. Hence the sequence (v_k) is orthonormal. By construction $\|(\lambda-A)P_k u_{n_{k+1}}\| \le \frac{1}{k}$ and consequently

$$\|(\lambda-A)v_{k+1}\| \le \|u_{n_{k+1}} - P_k u_{n_{k+1}}\|^{-1} \|(\lambda-A)(u_{n_{k+1}} - P_k u_{n_{k+1}})\|$$

$$\le \|(\lambda-A)u_{n_{k+1}}\| + \frac{1}{k} \qquad \text{for all } k,$$

so that $(\lambda-A)v_k \to 0$.

(3') ⇒ (4): Every sequence with the properties required in (3') has the properties required in (4).

(4) ⇒ (1): Assume (4) and $\lambda \notin \sigma_e(\overline{A})$. Then either $\lambda \in \rho(\overline{A})$ or λ is an isolated eigenvalue of finite multiplicity, and hence in any case there exists an $\varepsilon > 0$ such that $\dim\,(E(\lambda+\varepsilon)-E(\lambda-\varepsilon))H$ is finite. Take a sequence (u_n) according to (4). Since $u_n \rightharpoonup 0$ and $E(\lambda+\varepsilon)-E(\lambda-\varepsilon)$ is a compact operator, we have $(E(\lambda+\varepsilon)-E(\lambda-\varepsilon))u_n \to 0$. Put $v_n = u_n - (E(\lambda+\varepsilon)-E(\lambda-\varepsilon))u_n$. It follows $\liminf \|v_n\| > 0$ and

$$\|(\lambda-\overline{A})u_n\|^2 \ge \int_{-\infty}^{\lambda-\varepsilon} + \int_{\lambda+\varepsilon}^{\infty} |\lambda-t|^2 d\|E(t)u_n\|^2$$

$$= \int_{-\infty}^{\lambda-\varepsilon} + \int_{\lambda+\varepsilon}^{\infty} |\lambda-t|^2 d\|E(t)v_n\|^2 \ge \varepsilon^2 \|v_n\|^2,$$

hence $(\lambda-\overline{A})u_n \not\to 0$, a contradiction.

Theorems 1.1 and 1.2 in particular characterize the spectrum and the essential spectrum of the selfadjoint operator $\overline{A}$ in terms of its essentially selfadjoint restriction A. For convenience we shall write $\sigma(A)$ and $\sigma_e(A)$ instead of $\sigma(\overline{A})$ and $\sigma_e(\overline{A})$ respectively and we shall speak of the spectrum and essential spectrum of A.

2. Schrödinger operators

We consider a formal differential operator of second order defined by

$$Tu = \sum_{j=1}^{m} (i\partial_j + b_j)^2 u + qu$$

for all $u \in C_o^\infty(\mathbb{R}^m)$ (the set of infinitely differentiable complex functions of compact support on $\mathbb{R}^m$), where $\partial_j = \frac{\partial}{\partial x_j}$, and $b_1, \ldots, b_m, q$ are real measurable functions on $\mathbb{R}^m$ such that $|b_j|^4$ and $|q|^2$ are locally integrable and $\sum_{j=1}^{m} \partial_j b_j = 0$ holds in the sense of distributions (i.e. $\int \sum_{j=1}^{m} b_j \partial_j \varphi dx = 0$ for all $\varphi \in C_o^\infty(\mathbb{R}^m)$). The differentiation in the expression for Tu is understood in the sense of distributions. Under these conditions T obviously is a linear operator in $L_2(\mathbb{R}^m)$ with domain $D(T) = C_o^\infty(\mathbb{R}^m)$ and we have

$$Tu = -\Delta + 2i\sum_{j=1}^{m} b_j \partial_j u + b^2 u + qu,$$

with $b^2 = \sum_{j=1}^{m} b_j^2$, for every $u \in D(T)$. It follows that

$$\langle Tu|v\rangle - \langle u|Tv\rangle = -2i\int \sum_{j=1}^{m} b_j \partial_j (u^* v) dx = 0$$

for all $u,v \in D(T)$ (u^* denotes the complex conjugate of u, and we define the scalar product in $L_2(\mathbb{R}^m)$ by $\langle u|v\rangle = \int u^* v\, dx$). Since D(T) is dense, the operator T is symmetric.

The Schrödinger equation of a finite system of particles in a conservative field of force is of the form $i\partial_t u = Tu$ with an operator T such as described above; such operators are therefore called *Schrödinger* operators.

For applications it is most important to know conditions for the coefficients of T under which T is essentially selfadjoint. Such conditions have been given by T. Ikebe and T. Kato [10], they include some local conditions which we are going to state and a global condition.

For $\alpha > 0$ and a measurable function q on $\mathbb{R}^m$ define

$$M_{q,\alpha}(x) = \left\{ \int_{|x-y|\leq 1} |q(y)|^2 |x-y|^{4-m-\alpha} dy \right\}^{1/2}$$

and

$$N_q(x) = \left\{ \int_{|x-y|\leq 1} |q(y)|^2 dy \right\}^{1/2}$$

for $x \in \mathbb{R}^m$. Let $Q_{\alpha,loc}(\mathbb{R}^m)$ be the set of all measurable functions q such that $M_{q,\alpha}(x)$ is finite for all $x \in \mathbb{R}^m$ and the function $M_{q,\alpha}$ is bounded on every compact subset of $\mathbb{R}^m$. By $Q_\alpha(\mathbb{R}^m)$ we denote the set of all $q \in Q_{\alpha,loc}(\mathbb{R}^m)$ such that $M_{q,\alpha}$ is bounded on $\mathbb{R}^m$. For $q \in Q_{\alpha,loc}(\mathbb{R}^m)$ the function N_q is continuous, and for $q \in Q_\alpha(\mathbb{R}^m)$ it is bounded on $\mathbb{R}^m$.

2.1 Lemma. Let $\alpha > 0$ and $q^2 \in Q_{\alpha,loc}(\mathbb{R}^m)$. Then $q \in Q_{2+\beta,loc}(\mathbb{R}^m)$ for every $\beta \in]0,\frac{\alpha}{2}[$ and

$$[M_{q,2+\beta}(x)]^2 \leq CM_{q^2,\alpha}(x)$$

for all $x \in \mathbb{R}^m$ with $C > 0$ independent of q and x.

Proof. For $\beta \in]0,\frac{\alpha}{2}[$ we have

$$\left\{ \int_{|x-y|\leq 1} |q(y)|^2 |x-y|^{2-m-\beta} dy \right\}^2$$

$$\leq \int_{|x-y|\leq 1} |x-y|^{\alpha-2\beta-m} dy \int_{|x-y|\leq 1} |q(y)|^4 |x-y|^{4-m-\alpha} dy$$

$$= \frac{\omega_m}{\alpha-2\beta} \int_{|x-y|\leq 1} |q(y)|^4 |x-y|^{4-m-\alpha} dy$$

where ω_m is the surface of the unit sphere in $\mathbb{R}^m$, q.e.d.

The following local conditions for the coefficients of a Schrödinger operator will be assumed from now on:

2.2 Assumption. The functions $b_1, \ldots, b_m$ and q are real valued; for some $\alpha > 0$ we have

$$b^2 = \sum_{j=1}^{m} b_j^2 \in Q_{\alpha,\mathrm{loc}}(\mathbb{R}^m) \quad \text{and} \quad q \in Q_{\alpha,\mathrm{loc}}(\mathbb{R}^m);$$

the equation

$$\sum_{j=1}^{m} \partial_j b_j = 0$$

holds in the sense of distributions.

Note that this implies all conditions mentioned at the beginning of this section; hence T is a symmetric operator. Furthermore, by lemma 2.1, we have $b_j \in Q_{2+\beta,\mathrm{loc}}(\mathbb{R}^m)$ for $0 < \beta < \frac{\alpha}{2}$.

For the following estimates we use the notation

$$\|u\|_\Omega = \{\int_\Omega |u(x)|^2 dx\}^{1/2}$$

for open subsets Ω of $\mathbb{R}^m$. By Ω_ρ we denote the open set of all points of distance less than ρ from Ω.

2.3 Lemma. Let $\alpha > 0$ and $q \in Q_{\alpha,\mathrm{loc}}(\mathbb{R}^m)$. Then there exists a constant C (depending on m and α) such that

$$\|qu\|_\Omega \leq C\rho^{\alpha/2}\|M_{q,\alpha}\Delta u\|_{\Omega_\rho} + C\rho^{-m/2}\|N_q u\|_{\Omega_\rho}$$

for all $u \in C_o^\infty(\mathbb{R}^m)$, $\rho \in]0,1]$ and for all open sets $\Omega \subset \mathbb{R}^m$.

2.4 Lemma. Let $\beta > 0$ and $b_j \in Q_{2+\beta,\mathrm{loc}}(\mathbb{R}^m)$. Then there exists a constant C (depending on m and β) such that

$$\|\sum_{j=1}^{m} b_j \partial_j u\|_\Omega \leq C\rho^{\beta/2}\|M_{b,2+\beta}\Delta u\|_{\Omega_\rho} + C\rho^{-m/2-1}\|N_b u\|_{\Omega_\rho}$$

for all $u \in C_o^\infty(\mathbb{R}^m)$, $\rho \in]0,1]$ and for all open sets $\Omega \subset \mathbb{R}^m$ (with

$b = \{\sum_{j=1}^{m} b_j^2\}^{1/2}$).

The proof of the lemmas rests on the identity

$$u(x) = -\int k_\rho(|x-y|)\Delta u(y)\,dy + \int h_\rho(|x-y|)u(y)\,dy \tag{2.5}$$

valid for all $u \in C_o^\infty(\mathbb{R}^m)$ and all $\rho > 0$, where k_ρ and h_ρ are, for fixed $\rho > 0$, continuous real functions on $]0,\infty[$ vanishing identically on $[\rho,\infty[$, having piecewise contiuous derivatives and satisfying the inequalities (in case $m \geq 3$)

$$|k_\rho(r)| \leq Cr^{2-m}, \qquad |k_\rho'(r)| \leq Cr^{1-m}$$

$$|h_\rho(r)| \leq C\rho^{-m}, \qquad |h_\rho'(r)| \leq C\rho^{-m-1}$$

for $r \leq \rho$ with a constant C independent of r and ρ. Such functions have been constructed by F. Stummel [25]. We prove the lemmas for $m \geq 3$ only.

Proof of lemma 2.3. For $u \in C_o^\infty(\mathbb{R}^m)$ and $\rho \in]0,1]$ we estimate (2.5):

$$|u(x)|^2 \leq \Big\{C \int_{|x-y|\leq\rho} |x-y|^{2-m}|\Delta u(y)|\,dy +$$

$$+ C\rho^{-m} \int_{|x-y|\leq\rho} |u(y)|\,dy\Big\}^2$$

$$\leq 2C^2 \int_{|x-y|\leq\rho} |x-y|^{\alpha-m}dy \int_{|x-y|\leq\rho} |x-y|^{4-m-\alpha}|\Delta u(y)|^2 dy +$$

$$+ 2C^2\rho^{-2m} \int_{|x-y|\leq\rho} dy \int_{|x-y|\leq\rho} |u(y)|^2 dy$$

$$\leq C_1\rho^\alpha \int_{|x-y|\leq\rho} |x-y|^{4-m-\alpha}|\Delta u(y)|^2 dy +$$

$$+ C_1\rho^{-m} \int_{|x-y|\leq\rho} |u(y)|^2 dy \,.$$

Now multiply by $|q(x)|^2$, integrate over Ω and reverse the order of integrations; this gives

$$\|qu\|_{\Omega}^{2} \le C_1\rho^{\alpha}\int_{\Omega_\rho} |\Delta u(y)|^2 \int_{|x-y|\le 1} |q(x)|^2|x-y|^{4-m-\alpha}dx\, dy +$$

$$+C_1\rho^{-m}\int_{\Omega_\rho} |u(y)|^2 \int_{|x-y|\le 1} |q(x)|^2 dx\, dy$$

$$=C_1\rho^{\alpha}\|M_{q,\alpha}\Delta u\|_{\Omega_\rho}^{2} + C_1\rho^{-m}\|N_q u\|_{\Omega_\rho}^{2}, \qquad \text{q.e.d.}$$

<u>Proof of lemma 2.4</u>. Differentiate (2.5) to get

$$\partial_j u(x) = -\int k_\rho'(|x-y|)\,\frac{x_j-y_j}{|x-y|}\,\Delta u(y)\,dy+$$

$$+\int h_\rho'(|x-y|)\,\frac{x_j-y_j}{|x-y|}\,u(y)\,dy$$

and estimate as above. It follows

$$|\text{grad } u(x)|^2 \le m\Big\{C\int_{|x-y|\le\rho} |x-y|^{1-m}|\Delta u(y)|\,dy +$$

$$+ C\rho^{-m-1}\int_{|x-y|\le\rho} |u(y)|\,dy\Big\}^2$$

$$\le C_2\rho^{\beta}\int_{|x-y|\le\rho} |x-y|^{2-m-\beta}|\Delta u(y)|^2 dy + C_2\rho^{-m-2}\int_{|x-y|\le\rho} |u(y)|^2 dy.$$

Hence

$$\|\sum_{j=1}^{m} b_j\partial_j u\|_{\Omega}^{2} \le \int_{\Omega}|b(x)|^2|\text{grad } u(x)|^2 dx$$

$$\le C_2\rho^{\beta}\int_{\Omega_\rho}|\Delta u(y)|^2\int_{|x-y|\le 1}|b(x)|^2|x-y|^{2-m-\beta}dx\, dy +$$

$$+ C_2\rho^{-m-2}\int_{\Omega_\rho}|u(y)|^2\int_{|x-y|\le 1}|b(x)|^2 dx\, dy$$

$$= C_2\rho^{\beta}\|M_{b,2+\beta}\Delta u\|_{\Omega_\rho}^{2} + C_2\rho^{-m-2}\|N_b u\|_{\Omega_\rho}^{2}, \qquad \text{q.e.d.}$$

<u>2.6 Lemma.</u> Let $\beta > 0$ and $p \in Q_{2+\beta,\text{loc}}(\mathbb{R}^m)$. Then there exists a constant C (depending on m and β) such that

$$\|pu\|_{\Omega} \le C\rho^{\beta/2}\|M_{p,2+\beta}\text{ grad } u\|_{\Omega_\rho} + C\rho^{-m/2}\|N_p u\|_{\Omega_\rho}$$

for all $u \in C_o^\infty(\mathbb{R}^m)$, $\rho \in]0,1]$ and for all open sets $\Omega \subset \mathbb{R}^m$.

Proof. Integrating by parts in (2.5) we get

$$u(x) = -\int k_\rho'(|x-y|) \sum_{j=1}^m \frac{x_j - y_j}{|x-y|} \partial_j u(y)\, dy + \int h_\rho(|x-y|)u(y)\, dy$$

and hence the estimate

$$|u(x)|^2 \leq \Big\{ C \int_{|x-y|\leq\rho} |x-y|^{1-m} |\operatorname{grad} u(y)|\, dy + C\rho^{-m} \int_{|x-y|\leq\rho} |u(y)|\, dy \Big\}^2$$

$$\leq C_3\rho^\beta \int_{|x-y|\leq\rho} |x-y|^{2-m-\beta} |\operatorname{grad} u(y)|^2 dy + C_3\rho^{-m} \int_{|x-y|\leq\rho} |u(y)|^2 dy .$$

Multiplying by $|p(x)|^2$ and integrating we get

$$\|pu\|_\Omega^2 \leq C_3\rho^\beta \|M_{p,2+\beta} \operatorname{grad} u\|_{\Omega_\rho}^2 + C_3\rho^{-m}\|N_p u\|_{\Omega_\rho}^2 , \qquad \text{q.e.d.}$$

In an important special case we can use the lemmas to prove essential selfadjointness of T. We start from the most simple case $b_j = q = 0$: The operator $T = -\Delta$ is known to be essentially selfadjoint (see T. Kato [15], V. 5.2). Write the general operator, say S, in the form $S=T+A$ where $T = -\Delta$ and $A = 2i \sum_{j=1}^m b_j \partial_j + b^2 + q$. An operator A is called T-bounded, if $D(A) \supset D(T)$ and $\|Au\| \leq a\|u\| + b\|Tu\|$ for all $u \in D(T)$ with nonnegative constants a,b independent of u. The T-bound of A is the infimum of the set of numbers b such that the above inequality holds for some $a \geq 0$.

A well-known theorem of F. Rellich (see T. Kato [15], V.4.1) says that if T is essentially selfadjoint and A is symmetric and T-bounded with T-bound smaller than 1, then S=T+A is essentially selfadjoint, and its selfadjoint closure $\overline{S}$ is equal to $\overline{T+A}$ and has domain $D(\overline{S}) = D(\overline{T})$.

2.7 Theorem. Let $b_1,\ldots,b_m$ and q satisfy 2.2 and assume in addition $b^2 \in Q_\alpha(\mathbb{R}^m)$ and $q \in Q_\alpha(\mathbb{R}^m)$ for some $\alpha > 0$. Then $A = 2i \sum_{j=1}^m b_j \partial_j + b^2 + q$ with domain $D(A) = C_o^\infty(\mathbb{R}^m)$ is T-bounded ($T = -\Delta$) with T-bound zero.

Proof. Choose $\beta \in]0,\frac{\alpha}{2}[$; then according to lemma 2.1 we have $b \in Q_{2+\beta}(\mathbb{R}^m)$. The functions $M_{q,\alpha}, M_{b^2,\alpha}$ and $M_{b,2+\beta}$ are bounded; hence by lemma 2.3 and lemma 2.4 there exists a constant C_4 (independent of ρ and u) such that for all $u \in D(T)$ and $\rho \in]0,1]$ we have

$$\|qu\| \le C_4\rho^{\alpha/2}\|Tu\| + C_4\rho^{-m/2}\|N_q u\|$$

$$\|b^2u\| \le C_4\rho^{\alpha/2}\|Tu\| + C_4\rho^{-m/2}\|N_{b^2}u\|$$

$$\|\sum_{j=1}^{m} b_j\partial_j u\| \le C_4\rho^{\beta/2}\|Tu\| + C_4\rho^{-m/2-1}\|N_b u\|$$

and consequently

$$\|Au\| \le 4C_4\rho^{\beta/2}\|Tu\| + 4C_4\rho^{-m/2-1}\|Nu\| \tag{2.8}$$

with

$$N(x)=\{\int_{|x-y|\le 1}[|q(y)|^2+|b(y)|^2+|b(y)|^4]dy\}^{1/2}.$$

The function N ist bounded, say $N(x) \le C_5$ for all $x \in \mathbb{R}^m$. Given $\varepsilon > 0$ choose $\rho_o \in]0,1]$ such that $4C_4\rho_o^{\beta/2} \le \varepsilon$ and put $a(\varepsilon)=4C_4C_5\rho_o^{-m/2-1}$. Then by (2.8) we have $\|Au\| \le a(\varepsilon)\|u\|+\varepsilon\|Tu\|$ for all $u \in D(T)$, q.e.d.

2.9 Corollary. The Schrödinger operator S with coefficients $b_1,\ldots,b_m,q$ satisfying the assumptions of theorem 2.7 is essentially selfadjoint, and its selfadjoint extension $\overline{S}$ has domain $D(\overline{S}) = D(\overline{T})$ where $T = -\Delta$.

3. Perturbations small at infinity

In this section we consider perturbations $S=T+V$ of a Schrödinger operator T. We always assume that T is an essentially selfadjoint Schrödinger operator with domain $C_o^\infty(R^m)$ and that V is symmetric with $D(V)=D(T)$. We do not require, however, that S also is a Schrödinger operator in the sense of the definition given in section 2. This allows for nonlocal operators V (i.e. where the support of Vu is not always contained in the support of u, for $u \in C_o^\infty(R^m)$) such as integral operators or integro-differential operators. We shall give some examples in section 4. Our main concern is to establish relations between the essential spectra $\sigma_e(T)$ and $\sigma_e(S)$ (assuming S to be essentially selfadjoint) under certain conditions on V. The best known condition of this sort is T-*compactness* of V.

Let T and V denote operators in H. V is called T-*compact*, if $D(V) \supset D(T)$ and if for every bounded sequence (f_n) in $D(T)$ such that (Tf_n) is also bounded, the sequence (Vf_n) contains a convergent subsequence. This definition is more simply restated using the concept of the *Graph* $G(T)$ of T. Let $H \oplus H$ denote the Hilbert space of all ordered pairs (f,g) of elements of H with scalar product $\langle (f,g)|(f',g')\rangle = \langle f|f'\rangle + \langle g|g'\rangle$. $G(T)$ is the subspace of $H \oplus H$ consisting of all pairs (f,Tf) with $f \in D(T)$. If $D(T) \subset D(V)$, the mapping $(f,Tf) \to Vf$ is a linear operator L from $H \oplus H$ to H with domain $G(T)$. V is T-bounded iff L is bounded, and V is T-compact iff L is compact. It follows that if V is T-compact it is also T-bounded; but much more is true: If V is closable (i.e. V has a closed extension) and T-compact then V is T-bounded with T-bound zero (see corollary 3.4 below or S. Goldberg [5], corollary V.3.8). We also need the following fact: If V is T-bounded then for every sequence (f_n) in $D(T)$ such that $f_n \rightharpoonup 0$ and $Tf_n \rightharpoonup 0$ we have $Vf_n \rightharpoonup 0$. To prove this, note that L is bounded, hence L is continuous and consequently L is weakly continuous. This means that for every sequence (f_n, Tf_n) in $G(T)$ converging weakly to zero we have $Vf_n \rightharpoonup 0$; but (f_n, g_n) converges weakly to zero in $H \oplus H$ iff $f_n \rightharpoonup 0$ and $g_n \rightharpoonup 0$ in H.

3.1 Lemma. Let T and V be closable operators in Hilbert space such that $D(T) \subset D(V)$, and let $\overline{T}$ and $\overline{V}$ denote their closures (smallest closed extensions). Then the following statements are equivalent:

(1) V is T-compact,

(2) for every sequence (f_n) in $D(T)$ such that $f_n \rightharpoonup 0$ and $Tf_n \rightharpoonup 0$ we have $Vf_n \to 0$,

(3) $D(\overline{T}) \subset D(\overline{V})$ and for every sequence (f_n) in $D(\overline{T})$ such that $f_n \rightharpoonup 0$ and $\overline{T}f_n \rightharpoonup 0$ we have $\overline{V}f_n \to 0$,

(4) $\overline{V}$ is $\overline{T}$-compact

Proof. (1) $\Rightarrow$ (2): Let (f_n) be a sequence in $D(T)$ such that $f_n \rightharpoonup 0$ and $Tf_n \rightharpoonup 0$. Then, as shown above, $Vf_n \rightharpoonup 0$. Suppose that Vf_n does not converge strongly to zero. Then there is an $\varepsilon > 0$ and a subsequence (n_j) such that $\|Vf_{n_j}\| > \varepsilon$ for all j. Since the sequences (f_{n_j}) and (Tf_{n_j}) are bounded, the sequence (Vf_{n_j}) contains a convergent subsequence, say (Vf_{m_j}), by assumption (1). Since $Vf_{m_j} \rightharpoonup 0$ we also have $Vf_{m_j} \to 0$ contradicting $\|Vf_{m_j}\| > \varepsilon$.

(2) $\Rightarrow$ (3): From (2) it follows that for every sequence (f_n) in $D(T)$ such that $f_n \to 0$ and $Tf_n \to 0$ we have $Vf_n \to 0$, i.e. V is T-bounded. Therefore, for every sequence (f_n) in $D(T)$ such that (f_n) and (Tf_n) converge strongly, the sequence (Vf_n) also converges strongly; hence $D(\overline{T}) \subset D(\overline{V})$ and $\overline{V}$ is $\overline{T}$-bounded. Let (f_n) be a sequence in $D(\overline{T})$ such that $f_n \rightharpoonup 0$ and $\overline{T}f_n \rightharpoonup 0$. Then there is a sequence (g_n) in $D(T)$ such that $f_n - g_n \to 0$ and $\overline{T}(f_n - g_n) \to 0$. It follows that $\overline{V}(f_n - g_n) \to 0$, $g_n \rightharpoonup 0$ and $Tg_n \rightharpoonup 0$. By (2) we also have $Vg_n \to 0$ and hence $\overline{V}f_n = \overline{V}(f_n - g_n) + Vg_n \to 0$.

(3) $\Rightarrow$ (4): Let (f_n) be a bounded sequence in $D(\overline{T})$ such that $(\overline{T}f_n)$ is also bounded. Then there is a weakly convergent subsequence (f_{n_j})

such that $(\overline{T}f_{n_j})$ also converges weakly, say $f_{n_j} \rightharpoonup f$ and $\overline{T}f_{n_j} \rightharpoonup g$. The graph $G(\overline{T})$ is a closed subspace of $H \oplus H$, hence it is also weakly closed, and consequently $g=\overline{T}f$. But $f_{n_j}-f \rightharpoonup 0$ and $\overline{T}(f_{n_j}-f) \rightharpoonup 0$ imply $\overline{V}(f_{n_j}-f) \to 0$ by (3), i.e. $(\overline{V}f_{n_j})$ converges.

(4) $\Rightarrow$ (1): Since $D(T) \subset D(V)$ this is obvious.

We now specialize to operators in $L_2(\mathbb{R}^m)$:

<u>3.2 Theorem</u>. Let T and V be closable operators in $L_2(\mathbb{R}^m)$ such that $D(T) \subset D(V)$.

(1) If V is T-compact then for every $\varepsilon > 0$ there is a function $\varphi_\varepsilon \in C_o^\infty(\mathbb{R}^m)$ such that $\|Vu\| \leq \varepsilon(\|u\|+\|Tu\|)+\|\varphi_\varepsilon u\|$ for all $u \in D(T)$.

(2) If for every $\varphi \in C_o^\infty(\mathbb{R}^m)$ the operator $u \to \varphi u$ is T-compact and if the conclusion of (1) is true, then V is T-compact.

<u>Proof</u>. (1): If the assertion is not true, then there exists $\varepsilon > 0$ such that for every $\varphi \in C_o^\infty(\mathbb{R}^m)$ there is a function $u \in D(T)$ satisfying $\|u\|+\|Tu\|=1$ and $\|Vu\| > \varepsilon + \|\varphi u\|$. For every n choose $\varphi_n \in C_o^\infty(\mathbb{R}^m)$ such that $\varphi_n(x) \geq n$ for $|x| \leq n$ and choose $u_n \in D(T)$ such that $\|u_n\|+\|Tu_n\|=1$ and $\|Vu_n\| > \varepsilon + \|\varphi_n u_n\|$. Since the sequence (Vu_n) is bounded, we get

$$n^2 \int_{|x|\leq r} |u_n(x)|^2 dx \leq \|\varphi_n u_n\|^2 < \|Vu_n\|^2 \leq C$$

for all $n \geq r$. We also have $\|u_n\| \leq 1$ and consequently $u_n \rightharpoonup 0$. Assume now that $D(T)$ is dense: then T^* exists, $D(T^*)$ is dense, and for every $v \in D(T^*)$ we have $\langle v|Tu_n\rangle=\langle T^*v|u_n\rangle \to 0$. Since the sequence (Tu_n) is bounded, it follows $Tu_n \rightharpoonup 0$. If $D(T)$ is not dense, look at T as an operator from the Hilbert space $\overline{D(T)}$ into $L_2(\mathbb{R}^m)$ and reason as before. Hence $Tu_n \rightharpoonup 0$ in any case and consequently $Vu_n \to 0$ by lemma 3.1 contradicting $\|Vu_n\| > \varepsilon$.

(2) According to lemma 3.1 we have to show that $u_n \rightharpoonup 0$, $Tu_n \rightharpoonup 0$ im-

plies $Vu_n \to 0$. We may assume $\|u_n\|+\|Tu_n\| \leq 1$. Given $\varepsilon > 0$ there is a $\varphi_\varepsilon \in C_o^\infty(\mathbb{R}^m)$ such that $\|Vu_n\| \leq \varepsilon(\|u_n\|+\|Tu_n\|)+\|\varphi_\varepsilon u_n\|$ for all n and we have $\varphi_\varepsilon u_n \to 0$ by assumption and lemma 3.1. Hence $\limsup_{n\to\infty} \|Vu_n\| \leq \varepsilon$, which implies $Vu_n \to 0$.

3.3 Definition. Let T and V be operators in $L_2(\mathbb{R}^m)$. V is called *T-small at infinity* if V is T-bounded and if for every $\varepsilon > 0$ there exists $r(\varepsilon) > 0$ such that $\|Vu\| \leq \varepsilon(\|u\|+\|Tu\|)$ for all $u \in D(T)$ such that $u(x)=0$, a.e. in $\{x: x \in \mathbb{R}^m, |x| \leq r(\varepsilon)\}$.

3.4 Corollary. If T and V are closable operators in $L_2(\mathbb{R}^m)$ and if V is T-compact, then V is T-bounded with T-bound zero and V is T-small at infinity.

Proof. Apply part (1) of theorem 3.2. The first statement follows by estimating $\|\varphi_\varepsilon u\| \leq c_\varepsilon \|u\|$, the second is obtained with $r(\varepsilon) = \sup\{|x| : x \in \text{support } \varphi_\varepsilon\}$.

In order to apply theorem 3.2 to Schrödinger operators we need some lemmas. As in section 2 we always assume that the domain of the operator is $C_o^\infty(\mathbb{R}^m)$ and that the coefficients satisfy 2.2. We also recall the notation

$$\Omega_\rho = \{x \in \mathbb{R}^m : |x-y| < \rho \quad \text{for some } y \in \Omega\}.$$

3.5 Lemma. Let T be a Schrödinger operator in $L_2(\mathbb{R}^m)$. For every bounded open set $\Omega \subset \mathbb{R}^m$ and for every $\rho \in]0,1]$ there is a constant $C(\Omega,\rho)$ such that

$$\|\Delta u\|_\Omega \leq C(\Omega,\rho)\{\|u\|_{\Omega_\rho} + \|Tu\|_{\Omega_\rho}\}$$

for all $u \in D(T)$.

Proof. Choose $\beta \in]0,\frac{\alpha}{2}[$; then by lemma 2.1 we have

$$b \in Q_{2+\beta,loc}(\mathbb{R}^m) \quad \text{and} \quad p=|q|^{1/2} \in Q_{2+\beta,loc}(\mathbb{R}^m).$$

Given a bounded open set $\Omega \subset R^m$ and a number $\rho \in]0,1]$, the set Ω_ρ is also bounded and hence the functions

$$M_{q,\alpha}, N_q, M_{b^2,\alpha}, N_{b^2}, M_{b,2+\beta}, N_b, M_{p,2+\beta}, N_p$$

are all bounded on Ω_ρ; let C_o be a common upper bound.

Replace Ω, u, ρ by Ψ, v, σ respectively in lemmas 2.3, 2.4 and 2.6. It follows that for all open sets Ψ and numbers $\sigma \in]0,1]$ such that $\Psi_\sigma \subset \Omega_\rho$ and for all $v \in C_o^\infty(R^m)$ we have the estimates

$$\|qv\|_\Psi + \|b^2 v\|_\Psi \leq C_1(\sigma^{\alpha/2}\|\Delta v\|_{\Psi_\sigma} + \sigma^{-m/2}\|v\|_{\Psi_\sigma})$$

$$\|2\sum_{j=1}^m b_j \partial_j v\|_\Psi \leq C_1(\sigma^{\beta/2}\|\Delta v\|_{\Psi_\sigma} + \sigma^{-m/2-1}\|v\|_{\Psi_\sigma})$$

(3.6) $$\|bv\|_\Psi + \|pv\|_\Psi \leq C_1(\sigma^{\beta/2}\|\text{grad } v\|_{\Psi_\sigma} + \sigma^{-m/2}\|v\|_{\Psi_\sigma}),$$

where C_1 is $2CC_o$, C the constant in the lemmas. Consequently C_1 depends on Ω and ρ but not on v, Ψ or σ. Adding the first two inequalities we get

(3.7) $$\|Av\|_\Psi \leq 2C_1(\sigma^{\beta/2}\|\Delta v\|_{\Psi_\sigma} + \sigma^{-m/2-1}\|v\|_{\Psi_\sigma})$$

for

$$A = T + \Delta = 2i\sum_{j=1}^m b_j\partial_j + b^2 + q.$$

Let $\rho_o = \frac{1}{3}\rho$ and find a real function $\varphi \in C_o^\infty(R^m)$ with support in Ω_{ρ_o} and such that $\varphi(x) = 1$ for all $x \in \Omega$. For $u \in D(T)$ put $v = \varphi u$, $\Psi = \Omega_{\rho_o}$ and $\sigma < \rho_o$ in (3.7). Since the support of $v, \Delta v, Av$ is contained in Ψ this gives

$$\|Av\| \leq 2C_1(\sigma^{\beta/2}\|\Delta v\| + \sigma^{-m/2-1}\|v\|).$$

Now fix $\sigma_o < \rho_o$ such that $2C_1\sigma_o^{\beta/2} \leq \frac{1}{2}$; then

$$\|\Delta v\| \leq \|Tv\| + \|Av\| \leq \|Tv\| + \frac{1}{2}\|\Delta v\| + C_2\|v\|$$

with $C_2 = 2C_1\sigma_o^{-m/2-1}$ and consequently

$$
\begin{aligned}
\|\Delta u\|_{\Omega} \le \|\Delta v\| &\le 2\|T(\varphi u)\| + 2C_2\|\varphi u\| \\
&\le 2\|\varphi Tu\| + 4\|\operatorname{grad}\varphi \operatorname{grad} u\| + 2\|(\Delta\varphi)u\| + \\
&\quad + 4\|(\sum_{j=1}^{m} b_j \partial_j \varphi)u\| + 2C_2\|\varphi u\| \\
&\le C_3\{\|Tu\|_{\Omega_\rho} + \|\operatorname{grad} u\|_{\Omega_{\rho_o}} + \|bu\|_{\Omega_{\rho_o}} + \|u\|_{\Omega_\rho}\}
\end{aligned}
\tag{3.8}
$$

with a new constant C_3 depending on C_2 and upper bounds for φ and its derivatives, hence on Ω and ρ.

Choose a real function $\psi \in C_o^{\infty}(\mathbb{R}^m)$ with support in $\Omega_{2\rho_o}$ and such that $\psi(x)=1$ for $x \in \Omega_{\rho_o}$. Let $w=\psi u$; then

$$
\begin{aligned}
\|\psi \operatorname{grad} u\|^2 &= \int \psi^2 \sum_{j=1}^{m} |\partial_j u|^2 dx \\
&\le 2\int \psi^2 \{\sum_{j=1}^{m} |(i\partial_j + b_j)u|^2 + b^2|u|^2\} dx \\
&= 2\int u^* \sum_{j=1}^{m} (i\partial_j + b_j)[\psi^2(i\partial_j + b_j)u] dx + 2\|bw\|^2 \\
&= 2\int \{\psi^2 u^*(Tu - qu) + 2\psi u^* i\sum_{j=1}^{m} (\partial_j\psi)(i\partial_j + b_j)u\} dx + 2\|bw\|^2 \\
&= 2\langle w|\psi Tu\rangle - 2\langle w|qw\rangle + 4\langle u|i\sum_{j=1}^{m} (\partial_j\psi)(i\partial_j + b_j)w\rangle \\
&\quad + 4i\|u \operatorname{grad}\psi\|^2 + 2\|bw\|^2 \\
&\le \|w\|^2 + \|\psi Tu\|^2 + 2\|pw\|^2 + \frac{1}{32}\|\operatorname{grad} w\|^2 + 3\|bw\|^2 + C_4\|u\|^2_{\Omega_{2\rho_o}},
\end{aligned}
$$

where C_4 is an upper bound of $136|\operatorname{grad}\psi|^2$. From this estimate we get

$$
\begin{aligned}
\|\operatorname{grad} w\|^2 &\le 2\|\psi \operatorname{grad} u\|^2 + 2\|u \operatorname{grad}\psi\|^2 \\
&\le \frac{1}{16}\|\operatorname{grad} w\|^2 + 6\|bw\|^2 + 4\|pw\|^2 + C_5(\|Tu\|^2_{\Omega_\rho} + \|u\|^2_{\Omega_\rho})
\end{aligned}
$$

or

$$
\|\operatorname{grad} w\| + \|bw\| \le \frac{1}{4}\|\operatorname{grad} w\| + 4\|bw\| + 4\|pw\| + C_6(\|Tu\|_{\Omega_\rho} + \|u\|_{\Omega_\rho}). \tag{3.9}
$$

Now we use (3.6) with $\Psi=\Omega_{2\rho_o}$, $\sigma < \rho_o$ (so that $\Psi_\sigma \subset \Omega_{3\rho_o} = \Omega_\rho$) and v replaced by w; this yields

$$\|bw\| + \|pw\| \leq C_1(\sigma^{\beta/2}\|\text{grad } w\| + \sigma^{-m/2}\|w\|).$$

Choose σ_o so small that $4C_1\sigma_o^{\beta/2} \leq \frac{1}{4}$ (and $\sigma_o < \rho_o$); inserting this inequality in (3.9) we get

$$\|\text{grad } w\| + \|bw\| \leq C_7(\|Tu\|_{\Omega_\rho} + \|u\|_{\Omega_\rho})$$

with $C_7=2[C_6+4C_1\sigma_o^{-m/2}]$. Since $w(x) = u(x)$ for all $x \in \Omega_{\rho_o}$, we also have

$$\|\text{grad } u\|_{\Omega_{\rho_o}} + \|bu\|_{\Omega_{\rho_o}} \leq C_7(\|Tu\|_{\Omega_\rho} + \|u\|_{\Omega_\rho})$$

and the final result follows from (3.8).

<u>3.10 Lemma.</u> Let T be a Schrödinger operator in $L_2(R^m)$; then for every $\varphi \in C_o^\infty(R^m)$ the operator $u \to \varphi u$ is T-compact.

Proof. Given $\varphi \in C_o^\infty(R^m)$ we have to prove: For every bounded sequence (u_n) in D(T) such that the sequence (Tu_n) is also bounded, the sequence (φu_n) contains a convergent subsequence. Let $\Omega \subset R^m$ be a bounded open set containing the support of φ, and choose $\rho \in]0,1]$. Denote by v_n and Δv_n the restriction to Ω_ρ of u_n and Δu_n respectively and by w_n the restriction of u_n to Ω. By assumption and by lemma 3.5 (applied to Ω_ρ instead of Ω) the sequences (v_n) and (Δv_n) in $L_2(\Omega_\rho)$ are bounded. The identity (2.5) implies $w_n=-K_\rho \Delta v_n + H_\rho v_n$ where K_ρ and H_ρ are integral operators from $L_2(\Omega_\rho)$ into $L_2(\Omega)$ which are compact (see K. Jörgens [13], theorem 4.2). Hence the sequence (w_n) in $L_2(\Omega)$ contains a convergent subsequence (w_{n_j}); it follows that (φu_{n_j}) converges, q.e.d.

Combining theorem 3.2 with lemma 3.10 we get

3.11 Theorem. Let T be a Schrödinger operator in $L_2(\mathbb{R}^m)$ and V a closable operator with $D(V) \supset D(T)$. Then V is T-compact if and only if for every $\epsilon > 0$ there is a function $\varphi_\epsilon \in C_o^\infty(\mathbb{R}^m)$ such that

$$\|Vu\| \leq \epsilon(\|u\|+\|Tu\|) + \|\varphi_\epsilon u\|$$

for all $u \in D(T)$.

We give still another criterion for T-compactness which sometimes is more handy for applications:

3.12 Theorem. Let T be a Schrödinger operator in $L_2(\mathbb{R}^m)$ and V a closable operator with $D(V) \supset D(T)$. Then V is T-compact if and only if V is T-small at infinity and T-bounded with T-bound zero.

Proof. The "only if" part is corollary 3.4. To prove the "if" part, let $\epsilon > 0$ be given and choose $r > 0$ such that

$$\|Vu\| \leq \frac{\epsilon}{4} (\|u\|+\|Tu\|)$$

for all $u \in D(T)$ vanishing for $|x| \leq r$. Also choose a number $a > 0$ such that

$$\|Vu\| \leq a\|u\| + \frac{\epsilon}{4} \|Tu\|$$

for all $u \in D(T)$. Now construct a function $\varphi \in C_o^\infty(\mathbb{R}^m)$ such that $\varphi(x)=1$ for $|x| \leq r$ and $0 \leq \varphi(x) \leq 1$ for all $x \in \mathbb{R}^m$. Then $(1-\varphi)u$ vanishes in $|x| \leq r$ and hence

$$\begin{aligned} \|Vu\| &\leq \|V\varphi u\|+\|V(1-\varphi)u\| \\ &\leq a\|\varphi u\| + \frac{\epsilon}{4}\|T\varphi u\| + \frac{\epsilon}{4}(\|(1-\varphi)u\|+\|T(1-\varphi)u\|) \end{aligned} \tag{3.13}$$

for all $u \in D(T)$. Now we have

$$\begin{aligned} T\varphi u-\varphi Tu &= (1-\varphi)Tu - T(1-\varphi)u \\ &= 2 \operatorname{grad} \varphi \operatorname{grad} u + (\Delta\varphi)u + 2i(\sum_{j=1}^m b_j \partial_j \varphi)u \end{aligned}$$

and the norm of this expression can be estimatet with the help of lemmas 2.3 and 2.4 by

$$\eta\|\Delta u\|_{\Omega_1} + C(\eta)\|u\|_{\Omega_1}$$

where Ω is the support of φ and η may be taken arbitrarily small. Using lemma 3.5 this can be further estimated by

$$\eta\, C(\Omega_1,1)\|Tu\|_{\Omega_2} + (\eta C(\Omega_1,1)+C(\eta))\|u\|_{\Omega_2}.$$

We choose η such that $\eta C(\Omega_1,1)\leq 1$ and insert the estimate into (3.13); this gives

$$\begin{aligned}\|Vu\| &\leq a\|\varphi u\| + \tfrac{\varepsilon}{4}(\|\varphi Tu\|+\|Tu\|_{\Omega_2}) + \\ &\quad + \tfrac{\varepsilon}{4}(\|(1-\varphi)u\|+\|(1-\varphi)Tu\|+\|Tu\|_{\Omega_2})+C\|u\|_{\Omega_2} \\ &\leq \varepsilon(\|u\|+\|Tu\|) + C'\|u\|_{\Omega_2}.\end{aligned}$$

Finally if we choose $\varphi_\varepsilon \in C_o^\infty(R^m)$ such that $\varphi_\varepsilon(x) \geq C'$ for all $x \in \Omega_2$, we have $C'\|u\|_{\Omega_2} \leq \|\varphi_\varepsilon u\|$. By theorem 3.11 the proof is complete.

<u>3.14 Corollary.</u> Let $T=-\Delta$ with $D(T)=C_o^\infty(R^m)$ and

$$A = 2i\sum_{j=1}^{m} b_j\partial_j+b^2+q, \quad D(A) = D(T),$$

with $b_1,\ldots,b_m,q$ real, $b^2\in Q_\alpha(R^m)$, $q\in Q_\alpha(R^m)$ for some $\alpha>0$, $\sum_{j=1}^{m}\partial_j b_j=0$ in the sense of distributions and $N_{b^2}(x)\to 0$, $N_q(x)\to 0$ for $|x|\to\infty$. Then A is T-compact.

<u>Proof</u>. A is symmetric, and therefore it is closable. By theorem 2.7 A is T-bounded with T-bound zero. The estimate (2.8) holds for all $u \in D(T)$ and $\rho \in]0,1]$; the function N in this estimate tends to zero for $|x|\to\infty$ by assumption. Given $\varepsilon > 0$ choose $\rho \in]0,1]$ such that $4C_4\rho^{\beta/2} \leq \varepsilon$, then choose $r(\varepsilon) > 0$ such that $4C_4\rho^{-m/2-1}N(x) \leq \varepsilon$ for $|x|\geq r(\varepsilon)$. For all $u \in D(T)$ such that $u(x)=0$ for $|x|\leq r(\varepsilon)$, it fol-

lows $\|Au\| \leq \varepsilon(\|u\|+\|Tu\|)$. Hence A is T-small at infinity, and therefore A is T-compact by theorem 3.12.

More general criteria for T-compactness of a differential operator A of first order, where T is a Schrödinger operator, can be found in K. Jörgens [13].

If T is essentially selfadjoint, and V symmetric and T-compact, then V is closable and T-bounded with T-bound zero (corollary 3.4), hence S=T+V is essentially selfadjoint by Rellich's theorem. It is easy to see that T is S-bounded and it follows that V is S-compact. Combining the characterization 1.2 (4') of the essential spectrum with lemma 3.1 we get the well-known result that the essential spectra of T and S are equal. We now consider the more general case where V is T-small at infinity but not necessarily T-compact.

<u>3.15 Theorem.</u> Let T be an essentially selfadjoint Schrödinger operator in $L_2(R^m)$, and let V be symmetric and T-small at infinity. If T+V is essentially selfadjoint, then $\sigma_e(T) \subset \sigma_e(T+V)$; if in addition T is bounded below and if T is (T+V)-bounded, then

$$\min \{\lambda : \lambda \in \sigma_e(T)\} = \min \{\lambda : \lambda \in \sigma_e(T+V)\}.$$

Proof. Take $\lambda \in \sigma_e(T)$ and choose a sequence (u_n) in D(T) such that $\langle u_n, u_m \rangle = \delta_{nm}$ and $(\lambda-T)u_n \to 0$; this is possible according to 1.2 (3'). It suffices to prove $Vu_n \to 0$, because this implies $(\lambda-T-V)u_n \to 0$ and hence $\lambda \in \sigma_e(T+V)$ by 1.2(3'). By assumption there exists a > 0 such that

$$\|Vu\| \leq a(\|u\|+\|Tu\|) \tag{3.16}$$

for all $u \in D(T)$; also, given $\varepsilon > 0$, there exists r > 0 such that

$$\|Vu\| \leq \varepsilon(\|u\|+\|Tu\|) \tag{3.17}$$

for all $u \in D(T)$ vanishing for $|x| \leq r$. Now choose $\varphi \in C_o^\infty(R^m)$ as in the proof of theorem 3.12 and apply (3.16) and (3.17) to φu and $(1-\varphi)u$ respectively; also apply the estimate

(3.18) $$\|T\varphi u-\varphi Tu\| \leq C(\|u\|_{\Omega_2}+\|Tu\|_{\Omega_2})$$

used in that proof. This gives

$$\|Vu\| \leq \varepsilon(\|u\|+\|Tu\|) + C'(\|u\|_{\Omega_2}+\|Tu\|_{\Omega_2})$$

for all $u \in D(T)$ where $C'=a+(a+\varepsilon)C$. Apply this to u_n; we have $u_n \rightharpoonup 0$ and $Tu_n=\lambda u_n-(\lambda-T)u_n \rightharpoonup 0$ and hence $\|u_n\|_{\Omega_2}\to 0$ by lemma 3.10 and lemma 3.1. Moreover we have

$$\|Tu_n\|_{\Omega_2} \leq |\lambda|\,\|u_n\|_{\Omega_2} + \|(\lambda-T)u_n\| \to 0.$$

From the estimate above we get

$$\lim\sup \|Vu_n\| \leq \varepsilon \lim\sup (\|u_n\|+\|Tu_n\|).$$

Since $u_n \rightharpoonup 0$ and $Tu_n \rightharpoonup 0$ both sequences (u_n) and (Tu_n) are bounded, and hence $Vu_n \to 0$.

To prove the second statement let $\lambda_o = \min\{\lambda : \lambda \in \sigma_e(T)\}$; then $\min\{\lambda:\lambda \in \sigma_e(T+V)\} \leq \lambda_o$ because $\sigma_e(T) \subset \sigma_e(T+V)$. Consequently it suffices to prove $\lambda \geq \lambda_o$ for all $\lambda \in \sigma_e(T+V)$. Given $\lambda \in \sigma_e(T+V)$ choose a sequence (u_n) in $D(T)$ such that $\langle u_n|u_m\rangle = \delta_{nm}$ and $(\lambda-T-V)u_n \to 0$; this is possible according to 1.2(3'). It follows $u_n \rightharpoonup 0$, $(T+V)u_n \rightharpoonup 0$ and, since T is $(T+V)$-bounded, also $Tu_n \rightharpoonup 0$. Given $\varepsilon > 0$ choose $r > 0$ and $\varphi \in C_o^\infty(\mathbb{R}^m)$ as in the first part of the proof. By lemma 3.10 we have $\varphi u_n \to 0$ and hence $|\langle Vu_n|\varphi u_n\rangle| \leq \|Vu_n\|\|\varphi u_n\| \to 0$, the sequence (Vu_n) being bounded by (3.16). Since V is symmetric and $(1-\varphi)u_n$ vanishes for $|x|\leq r$ we have

$$\begin{aligned}|\langle Vu_n|(1-\varphi)u_n\rangle| &= |\langle u_n|V(1-\varphi)u_n\rangle| \\ &\leq \varepsilon\|u_n\|(\|(1-\varphi)u_n\|+\|T(1-\varphi)u_n\|) \\ &\leq \varepsilon\|u_n\|(1+C)(\|u_n\|+\|Tu_n\|)\end{aligned}$$

where we also have used (3.18). This gives

$$\limsup |\langle Vu_n|u_n\rangle|$$

$$\leq \limsup \{|\langle Vu_n|\varphi u_n\rangle| + |\langle Vu_n|(1-\varphi)u_n\rangle|\}$$

$$\leq \varepsilon(1+C) \limsup \|u_n\|(\|u_n\| + \|Tu_n\|)$$

and hence $\langle Vu_n|u_n\rangle \to 0$. Let $E(.)$ denote the spectral family of $\bar{T}$ (the selfadjoint extension of T). For every $\eta > 0$ we have $\dim E(\lambda_o-\eta)L_2(R^m) < \infty$ by definition of λ_o and because T is bounded below. From $u_n \rightharpoonup 0$ it follows $E(\lambda_o-\eta)u_n \to 0$ and therefore

$$\lambda = \lim \langle (T+V)u_n|u_n\rangle$$

$$= \lim \langle Tu_n|u_n\rangle$$

$$= \lim \langle Tu_n|(I-E(\lambda_o-\eta))u_n\rangle$$

$$\geq (\lambda_o-\eta) \lim \|(I-E(\lambda_o-\eta))u_n\|^2 = \lambda_o-\eta,$$

hence $\lambda \geq \lambda_o$, q.e.d.

<u>3.19 Corollary.</u> Let T be an essentially selfadjoint Schrödinger operator in $L_2(R^m)$, and let V be symmetric, T-small at infinity and such that T+V is essentially selfadjoint and T is (T+V)-bounded.
1) If T+V is a Schrödinger operator, then $\sigma_e(T)=\sigma_e(T+V)$.
2) If T is bounded below and $\sigma_e(T)=[\mu,\infty[$ for some $\mu \in R$, then $\sigma_e(T)=\sigma_e(T+V)$.

Proof. 1) Let S=T+V. By theorem 3.15 we have $\sigma_e(T) \subset \sigma_e(S)$. Since T is S-bounded, the assumptions of theorem 3.15 are also satisfied for S and -V; hence $\sigma_e(S) \subset \sigma_e(S-V)=\sigma_e(T)$. 2) By theorem 3.15 we have $[\mu,\infty[\subset \sigma_e(T+V)$ and $\min\{\lambda:\lambda \in \sigma_e(T+V)\} = \mu$, hence $\sigma_e(T+V)=[\mu,\infty[$.

<u>3.20 Remark.</u> If V is symmetric and T-bounded with T-bound smaller than 1, then T+V is essentially selfadjoint by Rellich's theorem and T is (T+V)-bounded. This clearly shows that in theorem 3.15 and corollary 3.19 the operator V needs not be T-compact.

An operator V in $L_2(R^m)$ with $D(V)=C_0^\infty(R^m)$ is called strictly local, if the support supp(Vu) is contained in supp(u) for every $u \in D(V)$, or equivalently if $\langle Vu|v\rangle = 0$ for all $u,v \in D(V)$ with $\text{supp}(u) \cap \text{supp}(v) = \emptyset$. Differential operators obviously are strictly local. We have not required in this section (except in part 1 of corollary 3.19) the perturbing operator V to be strictly local. However our assumptions imply a certain "approximate locality" as shown by the following theorem.

3.21 Theorem. Let T and V be operators in $L_2(R^m)$ such that V is symmetric and T-small at infinity. Then for every $\varepsilon > 0$ there is a number $\rho(\varepsilon) > 0$ such that

$$|\langle Vu|v\rangle| \leq \varepsilon(\|u\|^2+\|Tu\|^2+\|v\|^2+\|Tv\|^2)$$

for all $u,v \in D(T)$ such that supp(u) and supp(v) have a distance $d(\text{supp}(u), \text{supp}(v)) \geq \rho(\varepsilon)$ from each other.

Proof. Given $\varepsilon > 0$ there exists $r(\varepsilon) > 0$ such that $\|Vu\| \leq \varepsilon(\|u\|+\|Tu\|)$ for all $u \in D(T)$ such that $d(\{0\}, \text{supp}(u)) \geq r(\varepsilon)$. Suppose $d(\text{supp}(u),\text{supp}(v)) \geq \rho(\varepsilon) = 2r(\varepsilon)$; then either $d(\{0\},\text{supp}(u)) \geq r(\varepsilon)$ or $d(\{0\},\text{supp}(v)) \geq r(\varepsilon)$, hence either

$$\|Vu\| \leq \varepsilon(\|u\|+\|Tu\|)$$

or

$$\|Vv\| \leq \varepsilon(\|v\|+\|Tv\|).$$

In any case we have

$$|\langle Vu|v\rangle| = |\langle u|Vv\rangle| \leq \varepsilon(\|u\|+\|Tu\|)(\|v\|+\|Tv\|)$$

$$\leq \varepsilon(\|u\|^2+\|Tu\|^2+\|v\|^2+\|Tv\|^2).$$

4. Examples

As a first example we consider $T = -\Delta$ and a perturbing operator V defined by

$$Vu = \sum_{j,k=1}^{m} \partial_j a_{jk} \partial_k u$$

for $u \in C_o^\infty(R^m)$, where the a_{jk} are continuously differentiable functions satisfying $a_{jk}(x) = a_{kj}(x)^*$ for all $x \in R^m$. Then obviously V is a symmetric operator. If the functions a_{jk} and their derivatives are bounded, then V is T-bounded. To see this write

$$(4.1) \qquad Vu = \sum_{j,k=1}^{m} a_{jk} \partial_j \partial_k u + \sum_{j,k=1}^{m} (\partial_j a_{jk}) \partial_k u \text{ .}$$

The functions $b_k = \sum_{j=1}^{m} \partial_j a_{jk}$ are continuous and bounded, hence $b_k \in Q_{2+\beta}(R^m)$ for every $\beta \in \,]0,2[$. By lemma 2.4 the operator

$$u \to \sum_{j=1}^{m} b_j \partial_j u$$

is T-bounded with T-bound zero. Write $a^2 = \sum_{j,k=1}^{m} |a_{jk}|^2$ and $\alpha = \max\{a(x) : x \in R^m\}$ then

$$\|\sum_{j,k=1}^{m} a_{jk} \partial_j \partial_k u\|^2 = \int |\sum_{j,k=1}^{m} a_{jk} \partial_j \partial_k u|^2 dx$$

$$\leq \int a^2 \sum_{j,k=1}^{m} |\partial_j \partial_k u|^2 dx$$

$$(4.2) \qquad \leq \alpha^2 \sum_{j,k=1}^{m} \int |\partial_j \partial_k u|^2 dx$$

$$= \alpha^2 \sum_{j,k=1}^{m} \int (\partial_j^2 u)^* (\partial_k^2 u) dx$$

$$= \alpha^2 \int |\Delta u|^2 dx = \alpha^2 \|Tu\|^2 .$$

It follows that for every $\varepsilon > 0$ there is a constant $c(\varepsilon)$ such that $\|Vu\| \leq (\alpha + \varepsilon)\|Tu\| + c(\varepsilon)\|u\|$, i.e. V is T-bounded with T-bound not larger than α. Finally assume that the functions a_{jk} and their derivatives tend to zero for $|x| \to \infty$; then V is T-small at infinity. To prove this it suffices to show that both parts of V in (4.1) are T-small at infinity. For the first part this follows from (4.2) because α^2 can be replaced by $\max\{a(x)^2 : |x| \geq r\}$ in this estimate if $u(x) = 0$ for $|x| \leq r$; for the second part it follows from lemma 2.4 since $N_b(x) \to 0$ for $|x| \to \infty$. Now we have proved:

<u>4.3 Theorem.</u> Let $a_{jk} \in C^1(R^m)$, $a_{jk} = a_{kj}^*$,

$$a(x) = \Big\{ \sum_{j,k=1}^{m} |a_{jk}(x)|^2 \Big\}^{1/2} \to 0 \text{ and } b_k(x) = \sum_{j=1}^{m} \partial_j a_{jk}(x) \to 0$$

for $|x| \to \infty$, and $\alpha = \max\{a(x) : x \in R^m\}$. Then, for $T = -\Delta$, the operator

$$V = \sum_{j,k=1}^{m} \partial_j a_{jk} \partial_k \text{ with } D(V) = C_o^\infty(R^m)$$

is symmetric, T-bounded with T-bound not larger than α, and T-small at infinity.

<u>4.4. Corollary.</u> Under the assumptions of theorem 4.3 let in addition α be smaller than 1. Then T + V is essentially selfadjoint and $\sigma_e(T+V) = \sigma_e(T) = [0,\infty[$.

This follows at once from corollary 3.19 together with remark 3.20 and the observation that $T = -\Delta$ is bounded below (in fact $T \geq 0$) and has $\sigma_e(T) = [0,\infty[$ ($\overline{T}$ is unitarily equivalent to the maximal operator of multiplication by $|x|^2$ in $L_2(R^m)$).

The operator V is not T-compact (except for the case $V = 0$). To see this choose a unit vector $y \in R^m$ and a function $\varphi \in C_o^\infty(R^m)$ such that

$$g(x) = \sum_{j,k=1}^{m} a_{jk}(x) y_j y_k \varphi(x) \text{ is not identically zero,}$$

then choose $\varepsilon > 0$ such that $\varepsilon\|\varphi\| < \|g\|$. Define the sequence (u_n) in D(T) by

$$u_n(x) = \frac{1}{n^2} \varphi(x) e^{ixyn} .$$

Then obviously $u_n \to 0$, and by a simple calculation one finds $\|Tu_n\| \longrightarrow \|\varphi\|$ and $\|Vu_n\| \to \|g\|$. It follows that there does not exist a constant $a > 0$ such that $\|Vu\| \leq a\|u\|+\varepsilon\|Tu\|$ for all $u \in D(T)$, i.e. V does not have T-bound zero. By corollary 3.4 V is not T-compact.

Next we consider an integral operator

$$(4.5) \qquad Ku(x) = \int K(x,y)u(y)dy \ .$$

<u>4.6 Lemma.</u> Let $K(x,y) = K_1(x,y)\ K_2(x,y)$ where K_1 and K_2 are measurable functions on R^{2m} such that

$$\int |K_1(x,y)|^2 dy \leq C_1 \quad \text{for almost all} \quad x \in R^m$$

$$\int |K_2(x,y)|^2 dx \leq C_2 \quad \text{for almost all} \quad y \in R^m \ .$$

Then (4.5) defines a bounded operator K in $L_2(R^m)$ such that $\|K\|^2 \leq C_1 C_2$. If in addition $\int |K_2(x,y)|^2 dx \to 0$ for $|y| \to \infty$ then K is small at infinity, i.e. for every $\varepsilon > 0$ there exists $r(\varepsilon) > 0$ such that $\|Ku\| \leq \varepsilon\|u\|$ for all $u \in L_2(R^m)$ vanishing almost everywhere in the ball $|x| \leq r(\varepsilon)$.

<u>Proof.</u> The first part of the lemma is well known; we repeat the proof for the sake of the second part. For $u \in L_2(R^m)$ we have

$$|Ku(x)|^2 \leq \int |K_1(x,y)|^2 dy \int |K_2(x,y)|^2 |u(y)|^2 dy$$

$$\leq C_1 \int |K_2(x,y)|^2 u(y)|^2 dy$$

and consequently

$$\int |Ku(x)|^2 dx \leq C_1 \int |u(y)|^2 (\int |K_2(x,y)|^2 dx) dy \ .$$

The first assertion of the lemma follows easily; to prove the second, given $\varepsilon > 0$ choose $r(\varepsilon)$ such that $\int |K_2(x,y)|^2 dx \leq \frac{\varepsilon^2}{C_1}$ for $|y| \geq r(\varepsilon)$ and find $\int |Ku(x)|^2 dx \leq \varepsilon^2 \int |u(y)|^2 dy$, q.e.d.

Combining the lemma with theorem 3.12 we get:

<u>4.7 Corollary.</u> Under the assumptions of lemma 4.6 K is T-compact for every Schrödinger operator T in $L_2(R^m)$.

Now consider an integro-differential operator of the form

$$(4.8) \qquad Vu = \sum_{j=1}^{m} K_j \partial_j u, \quad u \in C_o^\infty(R^m)$$

where the K_j are integral operators as in lemma 4.6. For $u \in C_o^\infty(R^m)$ the inequality

$$(4.9) \qquad \|\partial_j u\|^2 \leq \|\text{grad } u\|^2 = - \langle u|\Delta u\rangle \leq \lambda^2\|\Delta u\|^2 + \frac{1}{4\lambda^2}\|u\|^2$$

holds for arbitrary $\lambda > 0$, hence

$$\|Vu\| \leq \sum_{j=1}^{m} \|K_j\| \, (\lambda\|\Delta u\| + \frac{1}{2\lambda}\|u\|)$$

which proves that V is T-bounded with T-bound zero for $T = -\Delta$. Similarly from the second statement of lemma 4.6 we derive that V is T-small at infinity. Combined with theorem 3.12 this gives:

<u>4.10 Corollary.</u> Let $K_1,\dots,K_m$ be integral operators satisfying all assumptions of lemma 4.6 and assume that the operator V defined by (4.8) is closable (e.g. symmetric). Then V is T-compact for $T = -\Delta$.

Finally we give an example of a (in general) non-local symmetric operator V which is not T-compact for $T = -\Delta$ but such that $\sigma_e(T+V) = \sigma_e(T)$: Let $a \in C^2(R^m)$ be a complex function such that a and its derivatives of first and second order tend to zero for $|x| \to \infty$, and denote by A the operator of multiplication by a in $L_2(R^m)$. Let K be a pseudo-differential operator of order 2 of the form

$$Kf(x) = \underset{n \to \infty}{\text{l.i.m.}} \, (2\pi)^{-m/2} \int_{|y|\leq n} k(y) e^{ixy} \hat{f}(y) dy$$

where $\hat{f}$ is the Fourier transform of f,

$$\hat{f}(y) = \underset{n \to \infty}{\text{l.i.m.}} \, (2\pi)^{-m/2} \int_{|y|\leq n} e^{-ixy} f(x) dx$$

and k is a real measurable function on R^m satisfying

$$(4.11) \qquad |k(y)| \leq 1+|y|^2 \text{ for all } y \in R^m \text{ and}$$

$$(4.12) \qquad \begin{array}{l} \text{there exists } z \in R^m \text{ with } |z| = 1 \text{ such that} \\ \lim_{n\to\infty} n^{-2} k(y+nz) = c \neq 0 \text{ for all } y \in R^m . \end{array}$$

K is defined on the subspace $D(K) = C_o^2(R^m)$ of all $f \in C^2(R^m)$ with compact support, because $(1+|y|^2)\hat{f}(y)$ is square integrable for such functions f, and because of (4.11).

<u>4.13 Theorem.</u> Let A,K be the operators defined above and let $V = A^*KA$ with $D(V) = C_o^\infty(R^m)$. Then V is symmetric, and for $T = -\Delta$ it is T-bounded with T-bound not exceeding α^2, where $\alpha = \max\{|a(x)| : x \in R^m\}$, and T-small at infinity. V is not T-compact (except for the case V = 0).

<u>4.14 Corollary.</u> Under the assumptions of theorem 4.13 let in addition α be smaller than 1. Then T + V is essentially selfadjoint and $\sigma_e(T+V) = \sigma_e(T) = [0,\infty[$.

The proof of the corollary is the same as for corollary 4.4.

<u>Proof of theorem</u> 4.13. The operator A has domain $L_2(R^m)$ and maps $C_o^\infty(R^m)$ into $C_o^2(R^m) = D(K)$; hence V is defined on $D(V) = C_o^\infty(R^m)$. For $f \in D(K)$ we have $(1+|y|^2)\hat{f}(y) = \widehat{(f-\Delta f)}(y)$; therefore, using (4.11), we get

$$\|Kf\| = \|k\hat{f}\| \leq \|f-\Delta f\| \leq \|f\| + \|\Delta f\|$$

and hence for $u \in D(T)$

$$\|Vu\| = \|a^*K(au)\| \leq \alpha\|K(au)\| \leq \alpha\|au\| + \alpha\ \|\Delta(au)\|$$

$$\leq \alpha(\|au\| + \|a\Delta\ u\| + 2\|\text{grad } a \text{ grad } u\| + \|u\Delta a\|).$$

Given $\varepsilon > 0$, by assumption there is a number $r_\varepsilon > 0$ such that $|a(x)| < \varepsilon/3\alpha$, $|\text{grad } a(x)| < \varepsilon/3\alpha$ and $|\Delta a(x)| < \varepsilon/3\alpha$ for $|x| \geq r_\varepsilon$. For $u \in D(T)$ such that $u(x) = 0$ for $|x| \leq r_\varepsilon$ this gives

$$\|Vu\| \leq \frac{\varepsilon}{3}\ (2\|u\| + \|Tu\| + 2\|\text{grad } u\|) \leq \varepsilon(\|u\| + \|Tu\|),$$

where we have also used (4.9) with $\lambda = 1$. Define

$$\alpha_1 = \max\{|\text{grad } a(x)| : x \in R^m\}$$

and

$$\alpha_2 = \max\{|\Delta a(x)| : x \in R^m\};$$

then for $u \in D(T)$ we get, using (4.9) again

$$\|Vu\| \leq \alpha(\alpha\|u\| + \alpha\|Tu\| + 2\alpha_1\|\text{grad } u\| + \alpha_2\|u\|)$$
$$\leq \alpha(\alpha+\alpha_2+\alpha_1/\lambda)\|u\| + \alpha(\alpha+2\alpha_1\lambda)\|Tu\|$$

for every $\lambda > 0$. Hence V is T-bounded with T-bound not larger than α^2 and T-small at infinity.

For $u \in D(T)$ we have

$$\langle u|Ku\rangle = (2\pi)^{-m/2} \int u(x)^* \int k(y)\, e^{ixy}\hat{u}(y)dydx$$
$$= \int k(y)\, |\hat{u}(y)|^2 dy$$

and this expression is real because k is real. Every $f \in D(K)$ is contained in $D(\overline{T}) = H_2^2(R^m)$; hence there is a sequence (u_j) in $D(T)$ such that $u_j \to f$ and $T\,u_j \to \overline{T}\,f$. Since K is T-bounded, we also have $K\,u_j \to K\,f$, which shows that $\langle f|Kf\rangle = \lim\langle u_j|Ku_j\rangle$ is real for all $f \in D(K)$, i.e. K is symmetric. It follows that $V = A^*KA$ is also symmetric.

Finally we prove that V (assuming $V \neq 0$) does not have T-bound zero; hence V is not T-compact by corollary 3.4. $V \neq 0$ implies $a \neq 0$; hence there exists $g \in D(T)$ with $\|g\| = 1$ such that $\||a|^2 g\| > 0$. Choose a vector $z \in R^m$ with $|z| = 1$ according to (4.12). Now define

$$u_n(x) = n^{-2}g(x)e^{inxz} \quad , \; x \in R^m, \; n=1,2,\ldots$$

Then obviously $u_n \in D(T)$, $u_n \to 0$, $\|Tu_n\| \longrightarrow \|g\| = 1$, and

$$Vu_n(x) = (2\pi)^{-m/2}\, n^{-2}\, a(x)^* \int k(y)e^{ixy}(\widehat{ag})(y-nz)dy$$
$$= (2\pi)^{-m/2} n^{-2} e^{inxz} a(x)^* \int k(y+nz)e^{ixy}(\widehat{ag})(y)dy$$

By (4.12) we have $n^{-2}k(y+nz) \to c$ for $n \to \infty$ and for every $y \in R^m$; by (4.11) the estimate

$$n^{-2}|k(y+nz)| \leq n^{-2}(1+|y+nz|^2)$$
$$\leq n^{-2}(1+2|y|^2+2n^2|z|^2) \leq 3+2|y|^2$$

holds for all $y \in R^m$ and $n=1,2,\ldots$ Since $(3+2|y|^2)(\widehat{ag})(y)$ is square integrable, $n^{-2}k(y+nz)(\widehat{ag})(y)$ tends to $c(\widehat{ag})(y)$ in square mean, and consequently

$$(2\pi)^{-m/2}\ n^{-2} \int k(y+nz)e^{ixy}(\widehat{ag})(y)dy$$

converges to cag in $L_2(R^m)$. This implies $\|Vu_n\| \to |c|\ \| |a|^2 g\| > 0$. If V has T-bound zero, then for every $\varepsilon > 0$ there exists $a_\varepsilon > 0$ such that

$$\|Vu\| \leq a_\varepsilon \|u\| + \varepsilon\ \|Tu\| \text{ for all } u \in D(T).$$

For $\varepsilon < |c|\ \| |a|^2 g\|$ and $u = u_n$ we get

$$|c|\ \| |a|^2 g\| = \lim \|Vu_n\| \leq \lim(a_\varepsilon \|u_n\| + \varepsilon \|Tu_n\|) = \varepsilon,$$

a contradiction. The proof is complete.

The operator V of theorem 4.13 is a differential operator iff k is a polynomial. On the other hand K, and hence V, is a strictly local operator iff K, and hence V, is a differential operator (see S. Helgason [7],p.242). It follows that V is strictly local iff k is a polynomial. More general examples with the same properties can be obtained by forming finite or even infinite sums of operators of this type.

5. Operators acting only on part of the variables

This section contains some preliminary results needed for the definition of Hamiltonian operators of N-particle systems to be given in section 6. We always consider operators in the Hilbert space $L_2(\mathbb{R}^{3N})$ and having domain $C_o^\infty(\mathbb{R}^{3N})$. For $x \in \mathbb{R}^{3N}$ we use the notation $x=(x_1,\dots,x_N)$ where $x_j \in \mathbb{R}^3$. Let $n \in \mathbb{N}$, $1 \leq n \leq N$ and $1 \leq j_1 < j_2 < \dots < j_n \leq N$. For brevity we write $x_{(1)}=(x_{j_1},\dots,x_{j_n})$, $x_{(2)}$ for the vector of the remaining variables (if $n < N$), $x=(x_{(1)},x_{(2)})$ and $u(x)=u(x_{(1)},x_{(2)})$ for functions u on $\mathbb{R}^{3N}$. Note that for $u \in C_o^\infty(\mathbb{R}^{3N})$ and for every $x_{(2)} \in \mathbb{R}^{3(N-n)}$ the function $u(\cdot,x_{(2)})$ is in $C_o^\infty(\mathbb{R}^{3n})$.

5.1 Definition. Let B denote an operator in $L_2(\mathbb{R}^{3N})$ with $D(B)=C_o^\infty(\mathbb{R}^{3N})$. B is said to act only on the variables $x_{(1)}=(x_{j_1},\dots,x_{j_n})$, if there exists an operator A in $L_2(\mathbb{R}^{3n})$ with $D(A)=C_o^\infty(\mathbb{R}^{3n})$ such that $(Bu)(x)=(Au(\cdot,x_{(2)}))(x_{(1)})$ for all $x \in \mathbb{R}^{3N}$ and for all $u \in C_o^\infty(\mathbb{R}^{3N})$; the operator A is then said to generate the operator B.

5.2 Theorem. For every operator A in $L_2(\mathbb{R}^{3n})$ with $D(A)=C_o^\infty(\mathbb{R}^{3n})$ and for every n-tuple of integers j_i such that $1 \leq j_1 < j_2 < \dots < j_n \leq N$ there exists a unique operator B in $L_2(\mathbb{R}^{3N})$ with $D(B)=C_o^\infty(\mathbb{R}^{3N})$ such that B acts only on the variables $x_{(1)}=(x_{j_1},\dots,x_{j_n})$ and is generated by A. A is symmetric iff B is symmetric; A is essentially selfadjoint iff B is essentially selfadjoint.

Proof. The operator B defined by $(Bu)(x)=(Au(\cdot,x_{(2)}))(x_{(1)})$ for

$x \in \mathbb{R}^{3N}$ and $u \in C_o^\infty(\mathbb{R}^{3N})$ has the required properties. If C is another operator with these properties, then by 5.1 we have $(Cu)(x) = (Au(\cdot,x_{(2)}))(x_{(1)}) = (Bu)(x)$ for all $x \in \mathbb{R}^{3N}$ and for all $u \in C_o^\infty(\mathbb{R}^{3N})$, and hence C=B.

Let A be symmetric, and denote by $\langle\cdot|\cdot\rangle_1$ and $\|\cdot\|_1$ the scalar product and the norm in $L_2(\mathbb{R}^{3n})$ respectively. We have

$$\langle Bu|v\rangle = \int\langle Au(\cdot,x_{(2)})|v(\cdot,x_{(2)})\rangle_1 dx_{(2)}$$

$$= \int\langle u(\cdot,x_{(2)})|Av(\cdot,x_{(2)})\rangle_1 dx_{(2)} = \langle u|Bv\rangle$$

for all $u,v \in C_o^\infty(\mathbb{R}^{3N})$. Since $C_o^\infty(\mathbb{R}^{3N})$ is dense, B is symmetric. If B is symmetric, the equation above holds for all $u,v \in C_o^\infty(\mathbb{R}^{3N})$. Put $u(x)=u_1(x_{(1)})u_2(x_{(2)})$ and $v(x)=v_1(x_{(1)})v_2(x_{(2)})$ where $u_1,v_1 \in C_o^\infty(\mathbb{R}^{3n})$ and $u_2,v_2 \in C_o^\infty(\mathbb{R}^{3(N-n)})$; if $\langle u_2|v_2\rangle_2 \neq 0$ this gives $\langle Au_1|v_1\rangle_1 = \langle u_1|Av_1\rangle_1$ for all $u_1,v_1 \in C_o^\infty(\mathbb{R}^{3n})$. Since $C_o^\infty(\mathbb{R}^{3n})$ is dense, A is symmetric.

Let A be essentially selfadjoint; then $(A-zI)D(A)$ is dense for every nonreal complex number z. Let $D_o(B)$ denote the subspace $C_o^\infty(\mathbb{R}^{3n}) \otimes C_o^\infty(\mathbb{R}^{3(N-n)})$ consisting of finite sums of elements $u=v \otimes w$ where $v \in D(A)$ and $w \in C_o^\infty(\mathbb{R}^{3(N-n)})$, i.e. $u(x)=v(x_{(1)})w(x_{(2)})$ for $x=(x_{(1)},x_{(2)}) \in \mathbb{R}^{3N}$. For these elements we have

$$(B-zI)(v \otimes w) = (A-zI)v \otimes w \quad \text{hence}$$

$$(B-zI)D_o(B) = (A-zI)D(A) \otimes C_o^\infty(\mathbb{R}^{3(N-n)}).$$

For nonreal z the subspace $(A-zI)D(A)$ of $L_2(\mathbb{R}^{3n})$ is dense, hence $(B-zI)D_o(B)$ is dense; consequently $(B-zI)D(B)$ is dense, which proves that B is essentially selfadjoint. If A is not essentially selfadjoint then for every $z \in \mathbb{C}\setminus\mathbb{R}$ there exists $f \in L_2(\mathbb{R}^{3n})$ such that $f\neq 0$ and $f \perp (A-zI)D(A)$. Hence for every $u \in D(B)$ and every $x_{(2)} \in \mathbb{R}^{3(N-n)}$ we have $f \perp (B-zI)u(\cdot,x_{(2)})$ and consequently $f\otimes g \perp (B-zI)D(B)$ for every $g \in L_2(\mathbb{R}^{3(N-n)})$. It follows that B is not essentially selfadjoint. This completes the proof.

We denote by $A_{j_1 \ldots j_n}$ the unique operator B existing according to theorem 5.2.

Let $\mu_1, \ldots, \mu_N$ be positive numbers. For every n-tuple of integers j_i such that $1 \leq j_1 < j_2 < \ldots < j_n \leq N$ define the operator $T^{j_1 \ldots j_n}$ in $L_2(\mathbb{R}^{3n})$ by

$$T^{j_1 \ldots j_n} u = - \sum_{i=1}^{n} (2\mu_{j_i})^{-1} \Delta_i u$$

for $u \in C_o^\infty(\mathbb{R}^{3n})$, where Δ_i denotes the Laplacean in the variables $x_i \in \mathbb{R}^3$. Apply the construction of theorem 5.2 to $A = T^{j_1 \ldots j_n}$; then $A_{j_1 \ldots j_n} = T^{j_1 \ldots j_n}_{j_1 \ldots j_n}$ is given by

$$T^{j_1 \ldots j_n}_{j_1 \ldots j_n} u = - \sum_{i=1}^{n} (2\mu_{j_i})^{-1} \Delta_{j_i} u$$

for $u \in C_o^\infty(\mathbb{R}^{3N})$. In the sequel we shall write $T_{j_1 \ldots j_n}$ instead of $T^{j_1 \ldots j_n}_{j_1 \ldots j_n}$ and T instead of $T_{1 \ldots N} = T^{1 \ldots N}_{1 \ldots N}$.

<u>5.3 Theorem.</u> $T_{j_1 \ldots j_n}$ is T-bounded with T-bound not larger than 1; more precisely we have $\|T_{j_1 \ldots j_n} u\| \leq \|Tu\|$ for all $u \in C_o^\infty(\mathbb{R}^{3N})$.

<u>Proof.</u> Let F denote the Fourier transformation in $L_2(\mathbb{R}^{3N})$. For $u \in C_o^\infty(\mathbb{R}^{3N})$ we have $(F\Delta_j u)(x) = -|x_j|^2 (Fu)(x)$ and hence

$$\|T_{j_1 \ldots j_n} u\|^2 = \int \Big\{ \sum_{i=1}^{n} (2\mu_{j_i})^{-1} |x_{j_i}|^2 |(Fu)(x)| \Big\}^2 dx$$

$$\leq \int \Big\{ \sum_{j=1}^{N} (2\mu_j)^{-1} |x_j|^2 |(Fu)(x)| \Big\}^2 dx = \|Tu\|^2 \qquad \text{q.e.d.}$$

5.4 Theorem. Suppose that A is an operator in $L_2(R^{3n})$ which is $T^{j_1\cdots j_n}$-bounded with $T^{j_1\cdots j_n}$-bound not exceeding b. Then $A_{j_1\cdots j_n}$ is $T_{j_1\cdots j_n}$-bounded and T-bounded with $T_{j_1\cdots j_n}$-bound and T-bound both not exceeding b.

Proof. By assumption there is, for every $\varepsilon > 0$, a number $a_\varepsilon \geq 0$ such that

$$\|Au\|_1 \leq a_\varepsilon\|u\|_1+(b+\varepsilon)\|T^{j_1\cdots j_n}u\|_1$$

for all $u \in C_o^\infty(R^{3n})$. It follows that for every $\delta > 0$ there is a number $c_\delta > 0$ such that

$$\|Au\|_1^2 \leq c_\delta^2\|u\|_1^2+(b+\delta)^2\|T^{j_1\cdots j_n}u\|_1^2$$

for all $u \in C_o^\infty(R^{3n})$. Consequently we have

$$\int|(A_{j_1\cdots j_n}u)(x_{(1)},x_{(2)})|^2dx_{(1)}=\|Au(\cdot,x_{(2)})\|_1^2$$

$$\leq c_\delta^2\|u(\cdot,x_{(2)})\|_1^2+(b+\delta)^2\|T^{j_1\cdots j_n}u(\cdot,x_{(2)})\|_1^2$$

$$= \int\{c_\delta^2|u(x_{(1)},x_{(2)})|_1^2+(b+\delta)^2|(T_{j_1\cdots j_n}u(x_{(1)},x_{(2)})|^2\}dx_{(1)}$$

for all $u \in C_o^\infty(R^{3N})$ and for all $x_{(2)}\in R^{3(N-n)}$. Integration with respect to $x_{(2)}$ gives

$$\|A_{j_1\cdots j_n}u\|^2 \leq c_\delta^2\|u\|^2+(b+\delta)^2\|T_{j_1\cdots j_n}u\|^2$$

which implies

$$\|A_{j_1\cdots j_n}u\| \leq c_\delta\|u\| + (b+\delta)\|T_{j_1\cdots j_n}u\|$$

for all $u \in C_o^\infty(R^{3N})$. Hence $A_{j_1\cdots j_n}$ is $T_{j_1\cdots j_n}$-bounded with $T_{j_1\cdots j_n}$-bound not exceeding b. By theorem 5.3 $T_{j_1\cdots j_n}$ can be replaced by T in the last inequality; hence $A_{j_1 \dots j_n}$ is also T-bounded with T-bound not exceeding b.

5.5 Theorem. Let A be $T^{j_1 \dots j_n}$-small at infinity. Then for every $\varepsilon > 0$ there exists a number $r_\varepsilon > 0$ such that

$$\|A_{j_1 \dots j_n} u\| \leq \varepsilon(\|u\| + \|T_{j_1 \dots j_n} u\|)$$

for all $u \in C_o^\infty(\mathbb{R}^{3N})$ such that $u(x)=0$ for all $x \in \mathbb{R}^{3N}$ with $(\sum_{i=1}^{n} |x_{j_i}|^2)^{1/2} \leq r_\varepsilon$.

Proof. By assumption and by definition 3.3, for every $\varepsilon > 0$ there is $r_\varepsilon > 0$ such that

$$\|Aw\|_1 \leq 2^{-1/2}\varepsilon(\|w\|_1 + \|T^{j_1 \dots j_n} w\|_1)$$

for all $w \in C_o^\infty(\mathbb{R}^{3n})$ satisfying $w(x)=0$ for $|x| \leq r_\varepsilon$; it follows that

$$\|Aw\|_1^2 \leq \varepsilon^2(\|w\|_1^2 + \|T^{j_1 \dots j_n} w\|_1^2)$$

for these functions w. Take $u \in C_o^\infty(\mathbb{R}^{3N})$ satisfying $u(x)=0$ for all $x \in \mathbb{R}^{3N}$ such that $(\sum_{i=1}^{n} |x_{j_i}|^2)^{1/2} \leq r_\varepsilon$; and let $w=u(\cdot, x_{(2)})$ for fixed $x_{(2)} \in \mathbb{R}^{3(N-n)}$. It follows that $w \in C_o^\infty(\mathbb{R}^{3n})$ and $w(x_{(1)})=0$ for $|x_{(1)}| \leq r_\varepsilon$, hence

$$\|(A_{j_1 \dots j_n} u)(\cdot, x_{(2)})\|_1^2 = \|Aw\|_1^2$$

$$\leq \varepsilon^2(\|w\|_1^2 + \|T^{j_1 \dots j_n} w\|_1^2)$$

$$= \varepsilon^2(\|u(\cdot, x_{(2)})\|_1^2 + \|(T_{j_1 \dots j_n} u)(\cdot, x_{(2)})\|_1^2).$$

Integration with respect to $x_{(2)}$ gives

$$\|A_{j_1 \dots j_n} u\|^2 \leq \varepsilon^2(\|u\|^2 + \|T_{j_1 \dots j_n} u\|^2),$$

and the assertion follows.

5.6 Theorem. Let A be symmetric and $T^{j_1 \cdots j_n}$-small at infinity. Then for every $\epsilon > 0$ there exists a number $\rho_\epsilon > 0$ such that

$$|\langle A_{j_1\cdots j_n} u|v\rangle| \le \epsilon\{\|u\|^2 + \|T_{j_1\cdots j_n} u\|^2 + \|v\|^2 + \|T_{j_1\cdots j_n} v\|^2\}$$

for all $u,v \in C_o^\infty(\mathbb{R}^{3N})$ such that the support of u and the support of v have distance $d(\operatorname{supp}(u), \operatorname{supp}(v)) > \rho_\epsilon$ from each other.

Proof. According to theorem 3.21, for every $\epsilon > 0$ there exists $\rho_\epsilon > 0$ such that

$$|\langle Au_1|v_1\rangle| \le \epsilon\{\|u_1\|_1^2 + \|T^{j_1\cdots j_n} u_1\|_1^2 + \|v_1\|_1^2 + \|T^{j_1\cdots j_n} v_1\|_1^2\}$$

for all $u_1, v_1 \in C_o^\infty(\mathbb{R}^{3n})$ satisfying $d(\operatorname{supp}(u_1), \operatorname{supp}(v_1)) \ge \rho_\epsilon$. Take $u,v \in C_o^\infty(\mathbb{R}^{3N})$ satisfying $d(\operatorname{supp}(u), \operatorname{supp}(v)) \ge \rho_\epsilon$ and let $u_1 = u(\cdot, x_{(2)})$, $v_1 = v(\cdot, x_{(2)})$ for fixed $x_{(2)} \in \mathbb{R}^{3(N-n)}$. Then $u_1, v_1 \in C_o^\infty(\mathbb{R}^{3n})$ and $d(\operatorname{supp}(u_1), \operatorname{supp}(v_1)) \ge \rho_\epsilon$, hence

$$|\langle (A_{j_1\cdots j_n} u)(\cdot, x_{(2)}) | v(\cdot, x_{(2)})\rangle_1| = |\langle Au_1|v_1\rangle|$$

$$\le \epsilon\{\|u_1\|_1^2 + \|T^{j_1\cdots j_n} u_1\|_1^2 + \|v_1\|_1^2 + \|T^{j_1\cdots j_n} v_1\|_1^2\}$$

$$= \epsilon\{\|u(\cdot, x_{(2)})\|_1^2 + \|(T_{j_1\cdots j_n} u)(\cdot, x_{(2)})\|_1^2 +$$

$$+ \|v(\cdot, x_{(2)})\|_1^2 + \|(T_{j_1\cdots j_n} u)(\cdot, x_{(2)})\|_1^2\}.$$

The assertion now follows by integration with respect to $x_{(2)}$.

6. N-particle Hamiltonians

The Hamiltonian operator (or Hamiltonian) of an N-particle system is of the form

$$(6.0) \qquad S=T+\sum_{n=1}^{N} \sum_{j_1<\dots<j_n} V_{j_1\dots j_n}+\sum_{n=2}^{N} \sum_{j_1<\dots<j_n} W_{j_1\dots j_n}$$

where T is the operator of kinetic energy of the particles, and for every n-tuple of integers j_i such that $1 \leq j_1< j_2<\dots< j_n \leq N$ the operators $V_{j_1\dots j_n}$ and $W_{j_1\dots j_n}$ represent an interaction of the subsystem consisting of the n particles numbered by $j_1,\dots,j_n$ with an external field of force and with each other respectively. We consider here the special case where all particles have spin zero; the generalization to the case of arbitrary spin will be indicated in the appendix. For spin zero S will be an operator in $L_2(R^{3N})$ with domain $D(S)=C_o^\infty(R^{3N})$. We proceed to discuss the operators T, $V_{j_1\dots j_n}$ and $W_{j_1\dots j_n}$.

The operator T of kinetic energy of the N particles has been defined in section 5; for j=1,2,...,N the positive number μ_j is the mass of the j-th particle. Similarly the operator $T_{j_1\dots j_n}$ defined in section 5 is the operator of the kinetic energy of the subsystem $j_1,\dots,j_n$.

6.1 Theorem. The operators $T_{j_1\dots j_n}$ and T are essentially selfadjoint.

Proof. By theorem 5.2 it suffices to show that the operator $T^{j_1\dots j_n}$ in $L_2(R^{3n})$ is essentially selfadjoint. Define the operator U in $L_2(R3n)$ by

$$(Uf)(x) = \prod_{i=1}^{n} (2\mu_{j_i})^{-3/4} f((2\mu_{j_1})^{-1/2}x_1,\ldots,(2\mu_{j_n})^{-1/2}x_n)$$

for all $x \in R^{3n}$ and $f \in L_2(R^{3n})$. It follows that U is unitary, $UC_o^\infty(R^{3n})=C_o^\infty(R^{3n})$, and $UT^{j_1\ldots j_n}U^*=-\Delta$ is the Laplace operator with domain $C_o^\infty(R^{3n})$. It is well known that $-\Delta$ in $L_2(R^{3n})$ with domain $C_o^\infty(R^{3n})$ is essentially selfadjoint (compare e.g. T.Kato [15],V.5.2). Hence $T^{j_1\ldots j_n}$ is unitarily equivalent to an essentially selfadjoint operator and is therefore essentially selfadjoint.

For every subsystem $j_1,\ldots,j_n$ of n particles ($1\leq n \leq N$) the operator $V_{j_1\ldots j_n}$ is supposed to be an operator acting only on the variables $x_{j_1},\ldots,x_{j_n}$ and which is generated by a symmetric operator $V^{j_1\ldots j_n}$ in $L_2(R^{3n})$ such that $V^{j_1\ldots j_n}$ is $T^{j_1\ldots j_n}$-small at infinity (note that by definition 3.3 this implies $T^{j_1\ldots j_n}$-boundedness of $V^{j_1\ldots j_n}$). The physical implication of these assumptions is that $V_{j_1\ldots j_n}$ represents an interaction of the subsystem $j_1,\ldots,j_n$ with an external field of force; in particular $T^{j_1\ldots j_n}$-smallness at infinity implies that the interaction is small whenever at least one of the particles of the subsystem has large distance from the origin. Such interactions are called n-<u>body-interactions with a field</u>. The theorems of section 5 imply the following

<u>6.2 Corollary.</u> a) $V_{j_1\ldots j_n}$ is symmetric and $T_{j_1\ldots j_n}$-bounded with $T_{j_1\ldots j_n}$-bound not exceeding the $T^{j_1\ldots j_n}$-bound of $V^{j_1\ldots j_n}$.

b) For every $\varepsilon > 0$ there exists a number $r_\varepsilon > 0$ such that

$$\|V_{j_1\ldots j_n}u\| \leq \varepsilon(\|u\|+\|T_{j_1\ldots j_n}u\|)$$

for all $u \in C_o^\infty(R^{3N})$ such that $u(x)=0$ for all x satisfying $(\sum_{i=1}^{n} |x_{j_i}|^2)^{1/2} \leq r_\varepsilon$.

c) For every $\varepsilon > 0$ there exists a number $\rho_\varepsilon > 0$ such that

$$|\langle V_{j_1\cdots j_n} u|v\rangle| \leq \varepsilon\{\|u\|^2+\|T_{j_1\cdots j_n}u\|^2+\|v\|^2+\|T_{j_1\cdots j_n}v\|^2\}$$

for all $u,v \in C_o^\infty(\mathbb{R}^{3N})$ such that $d(\mathrm{supp}(u),\mathrm{supp}(v)) \geq \rho_\varepsilon$.

For the definition of $W_{j_1\cdots j_n}$ we need the concept of internal coordinates of a subsystem: Let $j_1,\ldots,j_n$ be the subsystem $(2 \leq n \leq N)$, and denote by $j_{n+1},\ldots,j_N$ the rest of the integers $1,\ldots,N$ (if $n < N$). Let $a\colon \mathbb{R}^{3N} \to \mathbb{R}^{3N}$ be a linear mapping of the form $ax=(y_1,\ldots,y_N)$ where

$$(6.3)\qquad \begin{cases} y_i = \sum_{k=1}^{n} a_{ik}x_{j_k} & \text{for } i = 1,\ldots,n \\ y_i = x_{j_i} & \text{for } i = n+1,\ldots,N \end{cases}$$

and where the numbers a_{ik} satisfy

$$(6.3')\qquad a_{nk} = \mu^{-1}\mu_{j_k} \quad \text{for } k = 1,\ldots,n \quad \text{with } \mu=\sum_{k=1}^{n}\mu_{j_k}$$

and

$$(6.3'')\qquad \sum_{k=1}^{n}(2\mu_{j_k})^{-1}a_{ik}a_{jk}=(2\mu_i')^{-1}\delta_{ij} \quad \text{for } i,j = 1,\ldots,n$$

with positive numbers μ_i'. It follows that $\mu_n'=\mu$ whereas the positive numbers $\mu_1',\ldots,\mu_{n-1}'$ can be choosen arbitrarily; for convenience we require $\mu_i'=1/2$ for $i=1,\ldots,n-1$. Now μ is the total mass of the subsystem, and $y_n=\mu^{-1}\sum_{i=1}^{n}\mu_{j_i}x_{j_i}$ is the coordinate of its center of mass. The variables $y_1,\ldots,y_{n-1}$ are internal coordinates of the subsystem; they are invariant under translations of the entire system, i.e. if $x'=(x_1+z,\ldots,x_N+z)$ then

$$ax' = (y_1,\ldots,y_{n-1},y_n + z,\ldots,y_N + z).$$

To see this let $i=1,\ldots,n-1$ and $j=n$ in (6.3''); by (6.3') it follows

$\sum_{k=1}^{n} a_{ik}=0$, which proves the invariance of y_i under translations.

The equation (6.3'') also implies that the mapping a has determinant $|a|\neq 0$ and hence a is a bijection. Define a unitary operator U in $L_2(\mathbb{R}^{3N})$ by

(6.4) $$(Uf)(y)=|a|^{-1/2}f(a^{-1}y), \quad y \in \mathbb{R}^{3N}$$

for $f \in L_2(\mathbb{R}^{3N})$. Then we have $UC_o^\infty(\mathbb{R}^{3N})=C_o^\infty(\mathbb{R}^{3N})$, and the operator $T'=UTU^*$ has domain $C_o^\infty(\mathbb{R}^{3N})$ and can be decomposed as follows:

$$T'=T'_{1\ldots n-1}+T'_n+T'_{n+1\ldots N}$$

where the operators

$$T'_{1\ldots n-1}, T'_n \quad \text{and} \quad T'_{n+1\ldots N}$$

act only on the variables

$$(y_1,\ldots,y_{n-1}), y_n \quad \text{and} \quad (y_{n+1},\ldots,y_N)$$

and are generated by

$$-\sum_{i=1}^{n-1}\Delta_i, -(2\mu)^{-1}\Delta_n \quad \text{and} \quad -\sum_{i=n+1}^{N}(2\mu_{j_i})^{-1}\Delta_i$$

respectively. In internal coordinates $T'_{1\ldots n-1}$ is the operator of internal kinetic energy of the subsystem and T'_n represents the translation energy of the subsystem. The sum $T'_{1\ldots n}=T'_{1\ldots n-1}+T'_n$ represents the total kinetic energy of the subsystem; in fact we have $T'_{1\ldots n}=UT_{j_1\ldots j_n}U^*$. We now define the operators

(6.5) $$T_{j_1\ldots j_n}=U^*T'_{1\ldots n-1}U, \tilde{T}_{j_1\ldots j_n}=U^*T'_nU,$$

and we call them the operators of <u>internal kinetic energy</u> and of <u>translation energy</u> of the subsystem respectively. These operators are independent of the choice of internal coordinates: In fact an explicit calculation based on (6.3),(6.3') and (6.3'') shows that

(6.5')
$$\tilde{T}_{j_1\ldots j_n} = -(2\mu)^{-1} \sum_{i,k=1}^{n} \operatorname{grad}_{j_i} \operatorname{grad}_{j_k}$$

is independent of the transformation (6.3) and hence $\hat{T}_{j_1\ldots j_n} = T_{j_1\ldots j_n} - \tilde{T}_{j_1\ldots j_n}$ is also independent. In the special case n=N we put $\hat{T}_{1\ldots N} = \hat{T}$ and $\tilde{T}_{1\ldots N} = \tilde{T}$.

6.6 Lemma. The operators $\hat{T}_{j_1\ldots j_n}$ and $\tilde{T}_{j_1\ldots j_n}$ are essentially selfadjoint; the inequality

$$\|\hat{T}_{j_1\ldots j_n} u\| \leq \|T_{j_1\ldots j_n} u\|$$

holds for all $u \in C_o^\infty(R^{3N})$. If $j_1,\ldots,j_n$ is a subsystem of $k_1,\ldots,k_m$ ($n < m \leq N$) then

$$\|\hat{T}_{j_1\ldots j_n} u\| \leq \|\hat{T}_{k_1\ldots k_m} u\|$$

holds for all $u \in C_o^\infty(R^{3N})$.

Proof. By theorem 6.1 the operators $T'_{1\ldots n-1}$ and T'_n are essentially selfadjoint; as in theorem 5.3 we have

$$\|T'_{1\ldots n-1} u\| \leq \|T'_{1\ldots n} u\| \qquad \text{for all } u \in C_o^\infty(R^{3N}).$$

The first part of the assertion now follows from (6.5). To prove the second part choose internal coordinates $y_1,\ldots,y_{m-1}$ for the subsystem $k_1,\ldots k_m$ such that $y_1,\ldots,y_{n-1}$ are internal coordinates for $j_1,\ldots j_n$. In such coordinates the internal energy operators $T'_{1\ldots n-1}$ and $T'_{1\ldots m-1}$ of the two subsystems are generated by

$$-\sum_{i=1}^{n-1} \Delta_i \quad \text{and} \quad -\sum_{i=1}^{m-1} \Delta_i$$

respectively; it follows that

$$\|T'_{1\ldots n-1} u\| \leq \|T'_{1\ldots m-1} u\| \quad \text{for all } u \in C_o^\infty(R^{3N})$$

and hence

$$\|\hat{T}_{j_1\ldots j_n}u\| \leq \|\hat{T}_{k_1\ldots k_m}u\| \quad \text{for all } u \in C_o^\infty(R^{3N})$$

by unitary equivalence.

For every subsystem $j_1,\ldots,j_n$ of n particles ($2\leq n\leq N$) we define $W_{j_1\ldots j_n}$ in terms of internal coordinates and of the corresponding unitary operator given by (6.4):
$W_{j_1\ldots j_n}$ is assumed to be an operator with domain $C_o^\infty(R^{3N})$ such that $W'_{1\ldots n-1}=UW_{j_1\ldots j_n}U^*$ acts only on the internal coordinates $y_1,\ldots,y_{n-1}$ and is generated by a symmetric operator $W^{1\ldots n-1}$ in $L_2(R^{3n-3})$ which is Δ-small at infinity.

This description is independent of the choice of internal coordinates: Let $bx=(z_1,\ldots,z_N)$ be coordinates satisfying (6.3), (6.3') and (6.3'') with a_{ik} replaced by other real coefficients b_{ik} and with $2\mu_i'=1$ for $i=1,\ldots,n-1$; then the mapping $c=ba^{-1}$ is given by

$$z_i = \sum_{k=1}^{n-1} c_{ik}y_k \quad \text{for} \quad i=1,\ldots,n-1 \quad \text{and}$$

$$z_i = y_i \qquad \text{for} \quad i=n,\ldots,N$$

where (c_{ik}) is an orthogonal matrix of order n-1. If V is the unitary operator corresponding to b by (6.4) then

$$(UV^*f)(y) = f(cy) \quad \text{for} \quad y \in R^{3N} \quad \text{and} \quad f \in L_2(R^{3N});$$

hence $W'_{1\ldots n-1}$ acts only on the internal coordinates $y_1,\ldots,y_{n-1}$ iff $W''_{1\ldots n-1}=VW_{j_1\ldots j_n}V^*$ acts only on $z_1,\ldots,z_{n-1}$ and the generating operator of $W'_{1\ldots n-1}$ is Δ-small at infinity iff the same is true for the generating operator of $W''_{1\ldots n-1}$, moreover the Δ-bounds of these two operators are equal.

<u>6.7 Corollary.</u> a) $W_{j_1\ldots j_n}$ is symmetric and $\hat{T}_{j_1\ldots j_n}$-bounded with $T_{j_1\ldots j_n}$-bound not exceeding the Δ-bound of $W^{1\ldots n-1}$.

b) For every $\varepsilon > 0$ there exists a number $r_\varepsilon > 0$ such that

$$\|W_{j_1\ldots j_n}u\| \leq \varepsilon(\|u\|+\|\hat{T}_{j_1\ldots j_n}u\|)$$

for all $u \in C_o^\infty(\mathbb{R}^{3N})$ such that $u(x)=0$ for all x satisfying $(\sum_{i,k=1}^{n} |x_{j_i}-x_{j_k}|^2)^{1/2} \leq r_\varepsilon$.

c) For every $\varepsilon > 0$ there exists a number $\rho_\varepsilon > 0$ such that

$$|\langle W_{j_1\ldots j_n}u|v\rangle| \leq \varepsilon\{\|u\|^2+\|\hat{T}_{j_1\ldots j_n}u\|^2+\|v\|^2+\|\hat{T}_{j_1\ldots j_n}v\|^2\}$$

for all $u,v \in C_o^\infty(\mathbb{R}^{3N})$ such that $d(\mathrm{supp}(u),\mathrm{supp}(v)) \geq \rho_\varepsilon$.

Proof. a) $W^{1\ldots n-1}$ is symmetric, hence $W'_{1\ldots n-1}$ is symmetric by theorem 5.2, and consequently $W_{j_1\ldots j_n}=U^*W'_{1\ldots n-1}U$ is symmetric. $W^{1\ldots n-1}$ is Δ-bounded with Δ-bound γ, say, and $-\Delta$ generates $T'_{1\ldots n-1}$. Hence by theorem 5.4 $W'_{1\ldots n-1}$ is $T'_{1\ldots n-1}$-bounded with $T'_{1\ldots n-1}$-bound not larger than γ, and by (6.5) $W_{j_1\ldots j_n}$ is $\hat{T}_{j_1\ldots j_n}$-bounded with $\hat{T}_{j_1\ldots j_n}$-bound not exceeding γ.

b) $W^{1\ldots n-1}$ is Δ-small at infinity. By theorem 5.5, for every $\varepsilon > 0$ there exists $r'_\varepsilon > 0$ such that $\|W'_{1\ldots n-1}u\| \leq \varepsilon(\|u\|+\|T'_{1\ldots n-1}u\|)$ for all $u \in C_o^\infty(\mathbb{R}^{3N})$ such that $u(y)=0$ for all $y \in \mathbb{R}^{3N}$ with $(\sum_{i=1}^{n-1}|y_i|^2)^{1/2} \leq r'_\varepsilon$. Now put $u=Uv$ where $v \in C_o^\infty(\mathbb{R}^{3N})$; then by lemma 6.8 below there is a number $r_\varepsilon > 0$ such that $v(x)=0$ for $(\sum_{i,k=1}^{n}|x_{j_1}-x_{j_k}|^2)^{1/2} \leq r_\varepsilon$ implies $u(y)=0$ for $(\sum_{i=1}^{n-1}|y_i|^2)^{1/2} \leq r'_\varepsilon$. The assertion now follows from (6.5) and from the definition of $W_{j_1\ldots j_n}$.

c) By theorem 5.6, for every $\varepsilon > 0$ there exists $\rho'_\varepsilon > 0$ such that

$$|\langle W'_{1\ldots n-1}u|v\rangle| \leq \varepsilon\{\|u\|^2+\|T'_{1\ldots n-1}u\|^2+\|v\|^2+\|T'_{1\ldots n-1}v\|^2\}$$

for all $u,v \in C_o^\infty(\mathbb{R}^{3N})$ such that $d(\mathrm{supp}(u),\mathrm{supp}(v)) \geq \rho'_\varepsilon$. If $u=Uu_1$, $v=Uv_1$ then, by definition of U, there exists $\rho_\varepsilon > 0$ such that

$d(\mathrm{supp}(u_1),\mathrm{supp}(v_1)) \geq \rho_\varepsilon$ implies $d(\mathrm{supp}(u),\mathrm{supp}(v)) \geq \rho'_\varepsilon$. The assertion now follows from (6.5) and from the definition of $W_{j_1\ldots j_n}$.

<u>6.8 Lemma.</u> Let $y_1,\ldots,y_{n-1}$ be internal coordinates for the subsystem $j_1,\ldots,j_n$. Then there are positive numbers c_1,c_2 such that

$$c_1\Big(\sum_{i,k=1}^{n} |x_{j_i}-x_{j_k}|^2\Big)^{1/2} \leq \Big(\sum_{i=1}^{n-1} |y_i|^2\Big)^{1/2}$$

$$\leq c_2\Big(\sum_{i,k=1}^{n} |x_{j_i}-x_{j_k}|^2\Big)^{1/2}.$$

<u>Proof</u>. Let $y_1,\ldots,y_{n-1}$ be defined by (6.3); as mentioned before, (6.3') and (6.3'') imply $\sum_{k=1}^{n} a_{ik}=0$ for $i=1,\ldots,n-1$. It follows $y_i=\sum_{k=1}^{n-1} a_{ik}(x_{j_k}-x_{j_n})$ for $i=1,\ldots,n-1$ which implies the second inequality. Also, the matrix $(a_{ik})_{i,k=1,\ldots,n}$ being invertible, the submatrix $(a_{ik})_{i,k=1,\ldots,n-1}$ is invertible, and hence $x_{j_k}-x_{j_n}=\sum_{i=1}^{n-1} b_{ki}y_i$ for $k=1,\ldots,n-1$ with suitable coefficients b_{ki}. It follows that all terms $|x_{j_i}-x_{j_k}|$ can be estimated by a constant times $\Big(\sum_{i=1}^{n-1}|y_i|^2\Big)^{1/2}$, hence the result follows.

The physical implications of our assumptions on $W_{j_1\ldots j_n}$ are that this operator represents an interaction between the particles of the subsystem which is small whenever at least one of the relative distances of the particles is large. Such interactions are called <u>n-body-interactions</u>.

For every subsystem $k_1,\ldots,k_m$ $(2\leq m\leq N)$ we define the Hamiltonian and the internal Hamiltonian

(6.9) $$S_{k_1\ldots k_m}=T_{k_1\ldots k_m}+\sum_{(j_1\ldots j_n)\subset(k_1\ldots k_m)} (V_{j_1\ldots j_n}+W_{j_1\ldots j_n})$$

and

(6.10) $$\hat{S}_{k_1\ldots k_m}=\hat{T}_{k_1\ldots k_m}+\sum_{(j_1\ldots j_n)\subset(k_1\ldots k_m)} W_{j_1\ldots j_n}$$

respectively where $(j_1 \ldots j_n) \subset (k_1 \ldots k_m)$ denotes the inclusion relation for subsystems. For simplicity we have put $W_j=0$ for one-particle subsystems; also S_j is defined by $S_j=T_j+V_j$ for $j=1,\ldots,N$. The operators $\hat{S}_{k_1 \ldots k_m}$ are internal operators in the following sense:

6.11 Theorem. Let $x \mapsto ax=y$ be a coordinate transformation such that $y_1,\ldots,y_{m-1}$ are internal coordinates for the subsystem $k_1,\ldots,k_m$ $(2 \leq m \leq N)$, and let U denote the corresponding unitary operator defined by (6.4). Then the operator $U\hat{S}_{k_1 \ldots k_m}U^*$ acts only on the internal coordinates.

Proof. Since $U\hat{T}_{k_1 \ldots k_m}U^*$ acts only on the internal coordinates, it suffices to prove the same for $UW_{j_1 \ldots j_n}U^*$ and for every subsystem $j_1,\ldots,j_n$ of $k_1,\ldots,k_m$. Clearly it is enough to prove this for one system of internal coordinates. Choose coordinates $z=a_1x$ such that $z_1,\ldots,z_{n-1}$ are internal coordinates for the subsystem $j_1,\ldots,j_n$, and let $y=a_2z$ be such that $y_i=z_i$ for $i=1,\ldots,n-1$ and that $y_1,\ldots,y_{m-1}$ are internal coordinates for $k_1,\ldots,k_m$. If U_1 and U_2 are the corresponding unitary operators, then $U_1W_{j_1 \ldots j_n}U_1^*$ acts only on $z_1,\ldots,z_{n-1}$ by definition of $W_{j_1 \ldots j_n}$, and consequently

$$UW_{j_1 \ldots j_n}U^* = U_2U_1W_{j_1 \ldots j_n}U_1{}^*U_2{}^*$$

acts only on $y_1,\ldots,y_{n-1}$ by definition of U_2, q.e.d.

By theorem 6.1 and lemma 6.6 the operators $T_{k_1 \ldots k_m}$ and $\hat{T}_{k_1 \ldots k_m}$ are essentially selfadjoint. By corollary 6.2, lemma 6.6 and corollary 6.7 the $V_{j_1 \ldots j_n}$ are symmetric and $T_{k_1 \ldots k_m}$-bounded and the $W_{j_1 \ldots j_n}$ are symmetric and $\hat{T}_{k_1 \ldots k_m}$-bounded for $(j_1,\ldots,j_n) \subset (k_1,\ldots,k_m)$.

Our final assumption is:

(6.12) The operators $S_{k_1 \ldots k_m}$ and $\hat{S}_{k_1 \ldots k_m}$ are essentially selfadjoint and bounded below, $T_{k_1 \ldots k_m}$ is $S_{k_1 \ldots k_m}$-bounded and $\hat{T}_{k_1 \ldots k_m}$ is $\hat{S}_{k_1 \ldots k_m}$-bounded for every subsystem $k_1, \ldots, k_m$ of $1, \ldots, N$.

A sufficient condition for this is the following: Let $a_{j_1 \ldots j_n}$ and $b_{j_1 \ldots j_n}$ denote the $T_{j_1 \ldots j_n}$-bound of $V_{j_1 \ldots j_n}$ and the $\hat{T}_{j_1 \ldots j_n}$-bound of $W_{j_1 \ldots j_n}$ respectively ($b_j = 0$ for $j = 1, \ldots, N$); if

$$\sum_{(j_1 \ldots j_n) \subset (1 \ldots N)} (a_{j_1 \ldots j_n} + b_{j_1 \ldots j_n}) < 1$$

then the final assumption is true. In particular, this condition is fulfilled if $V_{j_1 \ldots j_n}$ is $T_{j_1 \ldots j_n}$-compact and if $W_{j_1 \ldots j_n}$ is $\hat{T}_{j_1 \ldots j_n}$-compact for all $j_1, \ldots, j_n$.

7. Symmetries of the Hamiltonian

Let γ be an orthogonal transformation in $\mathbb{R}^3$ (i.e. a rotation or a rotation followed by a reflection); define a unitary operator in $L_2(\mathbb{R}^{3N})$ by

$$(U(\gamma)f)(x) = f(\gamma^{-1}x_1, \gamma^{-1}x_2, \ldots, \gamma^{-1}x_N)$$

for $x \in \mathbb{R}^{3N}$ and $f \in L_2(\mathbb{R}^{3N})$. Obviously $U(\gamma)$ maps $C_o^\infty(\mathbb{R}^{3N})$ onto itself. An operator A in $L_2(\mathbb{R}^{3N})$ with domain $C_o^\infty(\mathbb{R}^{3N})$ is said to be *invariant under* γ if

$$U(\gamma)A = AU(\gamma).$$

From the definition of $T_{j_1 \ldots j_n}$ and T it follows that these operators are invariant under all orthogonal transformations γ; by (6.5') the same is true for the operators $\tilde{T}_{j_1 \ldots j_n}$ and hence for $\hat{T}_{j_1 \ldots j_n} = T_{j_1 \ldots j_n} - \tilde{T}_{j_1 \ldots j_n}$. An orthogonal transformation γ is said to be a *rotation symmetry* of the Hamiltonian S if the operators $S_{k_1 \ldots k_m}$ and $\hat{S}_{k_1 \ldots k_m}$ are invariant under γ for every subsystem $k_1, \ldots, k_m$; in particular γ is a rotation symmetry of S if all operators $V_{j_1 \ldots j_n}$ and $W_{j_1 \ldots j_n}$ in (6.0) are invariant under γ. The rotation symmetries of S obviously form a subgroup Γ of the three-dimensional orthogonal group $\mathfrak{O}(3)$; we call it the *rotation symmetry group* of S. Now $\mathfrak{O}(3)$ is a compact topological group. The mapping $\gamma \to U(\gamma)$ is continuous in the strong operator topology. Since the operators $S_{k_1 \ldots k_m}$ and $\hat{S}_{k_1 \ldots k_m}$

are symmetric they are closable; it follows that Γ is closed. Hence Γ is compact, and $\gamma \to U(\gamma)$ is a continuous representation of Γ by unitary operators in $L_2(R^{3N})$.

Let n_j be positive integers for $j=1,\dots,m$ such that $\sum_{j=1}^{m} n_j = N$; then

$$L_2(R^{3N}) = \hat{\bigotimes}_{j=1}^{m} L_2(R^{3n_j}) \quad \text{and} \quad U(\gamma) = \bigotimes_{j=1}^{m} U_j(\gamma) \quad {}^{*)}$$

when $U_j(\gamma)$ is defined by

$$(U_j(\gamma)f)(x) = f(\gamma^{-1}x_1,\dots,\gamma^{-1}x_{n_j})$$

for $x \in R^{3n_j}$ and $f \in L_2(R^{3n_j})$. The following lemmas on representations of compact groups obviously apply to this situation.

Let H_j for $j=1,\dots,m$ be Hilbertspaces and $H = \hat{\bigotimes}_{j=1}^{m} H_j$. For a compact group Γ, denote by D_α $(\alpha \in A)$ the continuous irreducible unitary (matrix) representations of Γ, and let U_j be a continuous representation of Γ by unitary operators in H_j for $j=1,\dots,m$. Then

$$U(\gamma) = \bigotimes_{j=1}^{m} U_j(\gamma) \qquad \text{for} \quad \gamma \in \Gamma$$

defines a continuous representation U of Γ by unitary operators in H. Let μ denote the unique invariant probability measure on Γ, and let d_α and χ_α be the dimension and the character of D_α respectively (i.e. $D_\alpha(\gamma) = (D_{ij}^{(\alpha)}(\gamma))$ is a complex $d_\alpha \times d_\alpha$-matrix and $\chi_\alpha(\gamma) = \sum_{j=1}^{d_\alpha} D_{jj}^{(\alpha)}(\gamma)$ for every $\gamma \in \Gamma$).

The operators

$$P_{j,\alpha} = d_\alpha \int_\Gamma \chi_\alpha(\gamma)^* U_j(\gamma)\, d\mu(\gamma), \qquad \alpha \in A,\ j=1,\dots,m$$

$$P_\alpha = d_\alpha \int_\Gamma \chi_\alpha(\gamma)^* U(\gamma)\, d\mu(\gamma), \qquad \alpha \in A$$

---

*) By $H_1 \hat{\otimes} H_2$ we denote the completion of $H_1 \otimes H_2$. For operators A,B,C in $H_1 \hat{\otimes} H_2$, H_1 and H_2 respectively the equation $A = B \otimes C$ means: $f \in D(B)$, $g \in D(C)$ imply $f \otimes g \in D(A)$ and $A(f \otimes g) = Bf \otimes Cg$.

are orthogonal projections in H_j and in H respectively satisfying

(7.0) $$P_{j,\alpha}P_{j,\beta}=\delta_{\alpha\beta}P_{j,\alpha},\quad P_\alpha P_\beta=\delta_{\alpha\beta}P_\alpha,\quad \alpha,\beta \in A,\quad j=1,\dots,m$$

and

(7.0') $$\sum_{\alpha\in A} P_{j,\alpha}=I_j,\qquad \sum_{\alpha\in A} P_\alpha=I$$

where I_j and I are the identity in H_j and H respectively. We also need the equations

(7.0") $$d_\alpha\int_\Gamma D_{kh}^{(\alpha)}(\gamma)^*D_{ij}^{(\beta)}(\gamma)\,d\mu(\gamma) = \delta_{\alpha\beta}\delta_{ki}\delta_{hj}$$

(compare M.A. Neumark [21], II.6 for these concepts and results).

For $\alpha \in A$ we define closed subspaces $H_{j,\alpha}=P_{j,\alpha}H_j$ of H_j and $H_\alpha=P_\alpha H$ of H. If $\alpha \neq \beta$ then $P_\alpha P_\beta=0$ and consequently H_α and H_β are orthogonal; similarly $H_{j,\alpha}$ and $H_{j,\beta}$ are orthogonal for $\alpha \neq \beta$ and for $j=1,\dots,m$. By definition of $P_{j,\alpha}$ and P_α we have

$$P_{j,\alpha}U_j(\gamma)=U_j(\gamma)P_{j,\alpha}\quad \text{and}\quad P_\alpha U(\gamma)=U(\gamma)P_\alpha$$

for all $\alpha \in A$, $j=1,\dots,m$ and for all $\gamma \in \Gamma$. It follows that the restriction $U_j(\gamma)|H_{j,\alpha}$ is a unitary operator in $H_{j,\alpha}$ and similarly $U(\gamma)|H_\alpha$ is a unitary operator in H_α for all $\gamma \in \Gamma$.

A closed subspace N of H_j (or H) is called D_α-generating, if $U_j(\gamma)|N$ (or $U(\gamma)|N$) is a unitary operator in N for all $\gamma \in \Gamma$ and if there exists an orthonormal basis of N such that for every $\gamma \in \Gamma$ the operator $U_j(\gamma)|N$ (or $U(\gamma)|N$) is represented by the matrix $D_\alpha(\gamma)$ with respect to this basis. Consequently, every D_α-generating subspace is finite-dimensional and its dimension is equal to the dimension d_α of D_α.

<u>7.1 Lemma.</u> (1) Every D_α-generating subspace of H_j and H is contained in $H_{j,\alpha}$ and H_α respectively.

(2) The subspaces $H_{j,\alpha}$ and H_α are orthogonal sums of D_α-generating subspaces of H_j and H respectively.

Proof. It suffices to prove the statements concerning H:
(1) Let N be a D_α-generating subspace of H and let $\{u_1,\ldots,u_{d_\alpha}\}$ be an orthonormal basis of N such that for every $\gamma\in\Gamma$ we have

$$U(\gamma)u_j = \sum_{i=1}^{d_\alpha} D_{ij}^{(\alpha)}(\gamma)u_i \quad \text{for} \quad j=1,\ldots,m.$$

From the definition of P_α and from the identity (7.0") it follows

$$P_\alpha u_j = d_\alpha \int_\Gamma \chi_\alpha(\gamma)^* \sum_{i=1}^{d_\alpha} D_{ij}^{(\alpha)}(\gamma)u_i\, d\mu(\gamma)$$

$$= \sum_{k,i=1}^{d_\alpha} d_\alpha \int_\Gamma D_{kk}^{(\alpha)}(\gamma)^* D_{ij}^{(\alpha)}(\gamma)\, d\mu(\gamma)u_i = u_j.$$

Hence $u_j\in H_\alpha$ for j=1,...,m, which proves $N\subset H_\alpha$.

(2) For every $\gamma\in\Gamma$, $U(\gamma)|H_\alpha$ is an unitary operator in H_α, hence $\gamma\mapsto U(\gamma)|H_\alpha$ is a continuous representation of Γ by unitary operators in H_α. Every such representation is the direct sum of finite-dimensional irreducible representations (see K. Maurin [20], IV.3), i.e. H_α is the orthogonal sum of D_β-generating subspaces for $\beta\in B$, where B is a subset of A. By part (1) of the lemma every D_β-generating subspace of H is contained in H_β, and H_β is orthogonal to H_α for $\alpha\neq\beta$. Hence $B=\{\alpha\}$, q.e.d.

For $\alpha_1,\alpha_2,\ldots,\alpha_m\in A$ define

$$D(\gamma) = \bigotimes_{j=1}^{m} D_{\alpha_j}(\gamma) \quad \text{for every } \gamma\in\Gamma; \tag{7.2}$$

then the mapping $\gamma\mapsto D(\gamma)$ is a continuous unitary representation of Γ and is therefore unitarily equivalent to a direct sum of irreducible unitary representations, say

$$D \cong \bigoplus_{\alpha\in A(\alpha_1,\ldots,\alpha_m)} n_\alpha D_\alpha, \tag{7.2'}$$

where $A(\alpha_1,\dots,\alpha_m)$ is a finite subset of A, and where for every $\alpha \in A(\alpha_1,\dots,\alpha_m)$ the positive integer n_α indicates how many times the representation D_α has to be repeated in the sum. We shall often write $(\alpha_1,\dots,\alpha_m)<\alpha$ instead of $\alpha \in A(\alpha_1,\dots,\alpha_m)$.

7.3 Lemma. Let N_j be a D_{α_j}-generating subspace of H_j for $j=1,\dots,m$. Then for every $\alpha \in A(\alpha_1,\dots,\alpha_m)$ there exist n_α mutually orthogonal D_α-generating subspaces $M_{\alpha i}$ of H such that

$$\bigotimes_{j=1}^{m} N_j = \bigoplus_{\alpha\in A(\alpha_1,\dots,\alpha_m)} \bigoplus_{i=1}^{n_\alpha} M_{\alpha i}.$$

Proof. Let $N=\bigotimes_{j=1}^{m} N_j$; by assumption $U(\gamma)|N$ is a unitary operator in N for every $\gamma \in \Gamma$, and there exists an orthonormal basis of N such that for every $\gamma \in \Gamma$ the operator $U(\gamma)|N$ is represented by the matrix $D(\gamma)=\bigotimes_{j=1}^{m} D_{\alpha_j}(\gamma)$ with respect to this basis. According to (7.2') there exists another orthonormal basis of N which decomposes N into orthogonal subspaces $M_{\alpha i}$ $(i=1,\dots,n_\alpha;\ \alpha \in A(\alpha_1,\dots,\alpha_m))$ such that $M_{\alpha i}$ is D_α-generating, q.e.d.

7.4 Lemma. For $\alpha_1,\dots,\alpha_m \in A$ the inclusion

$$\hat{\bigotimes}_{j=1}^{m} H_{j,\alpha_j} \subset \bigoplus_{\alpha\in A(\alpha_1,\dots,\alpha_m)} H_\alpha$$

holds.

Proof. By lemma 7.1(2) the space $\hat{\bigotimes}_{j=1}^{m} H_{j,\alpha_j}$ is an orthogonal sum of subspaces $\bigotimes_{j=1}^{m} N_j$ where N_j is D_{α_j}-generating for $j=1,\dots,m$. By lemma 7.3 and by lemma 7.1(1) we have

$$\bigotimes_{j=1}^{m} N_j \subset \bigoplus_{\alpha\in A(\alpha_1,\dots,\alpha_m)} H_\alpha, \quad \text{q.e.d.}$$

7.5 Lemma. For every $\alpha_o \in A$ the inclusion

$$H_{\alpha_o} \subset \bigoplus_{(\alpha_1,\ldots,\alpha_m)<\alpha_o} (\hat{\bigotimes}_{j=1}^{m} H_{j,\alpha_j})$$

holds.

Proof. From (7.0') we get

$$\bigoplus_{\alpha\in A} H_{j,\alpha} = H_j \quad \text{and} \quad \bigoplus_{\alpha\in A} H_\alpha = H,$$

hence

$$H = \bigoplus_{\alpha_1,\ldots,\alpha_m \in A} (\hat{\bigotimes}_{j=1}^{m} H_{j,\alpha_j})$$

$$= [\bigoplus_{(\alpha_1,\ldots,\alpha_m)<\alpha_o} (\hat{\bigotimes}_{j=1}^{m} H_{j,\alpha_j})] \oplus [\bigoplus_{(\alpha_1,\ldots,\alpha_m)\nless\alpha_o} (\hat{\bigotimes}_{j=1}^{m} H_{j,\alpha_j})].$$

By lemma 7.4 $\hat{\bigotimes}_{j=1}^{m} H_{j,\alpha_j}$ is orthogonal to H_{α_o} if $(\alpha_1,\ldots,\alpha_m)\nless\alpha_o$; this proves the lemma.

The following lemma refers to the relation $(\alpha_1,\ldots,\alpha_m)<\alpha$ which has been defined by (7.2) and (7.2') for arbitrary $m > 1$:

7.6 Lemma. If $(\alpha_1,\ldots,\alpha_{j-1},\alpha_{j_1},\alpha_{j_2},\alpha_{j+1},\ldots,\alpha_m)<\alpha_o$ then there exists at least one $\alpha_j \in A$ such that

$$(\alpha_1,\ldots,\alpha_{j-1},\alpha_j,\alpha_{j+1},\ldots,\alpha_m)<\alpha_o \quad \text{and} \quad (\alpha_{j_1},\alpha_{j_2})<\alpha_j.$$

Proof. Denote by $D_{\alpha_1}\otimes\ldots\otimes D_{\alpha_m}$ the representation of Γ defined by (7.2). By assumption

$$D = D_{\alpha_1}\otimes\ldots\otimes D_{\alpha_{j-1}}\otimes D_{\alpha_{j_1}}\otimes D_{\alpha_{j_2}}\otimes D_{\alpha_{j+1}}\otimes\ldots\otimes D_{\alpha_m}$$

is unitarily equivalent to a direct sum of irreducible representations in which D_{α_o} occurs. By (7.2') we also have

$$D_{\alpha_{j_1}} \otimes D_{\alpha_{j_2}} \cong \bigoplus_{\beta \in A(\alpha_{j_1},\alpha_{j_2})} n_\beta D_\beta;$$

hence D is also unitarily equivalent to a direct sum of representations

$$D_{\alpha_1} \otimes \ldots \otimes D_{\alpha_{j-1}} \otimes D_\beta \otimes D_{\alpha_{j+1}} \otimes \ldots \otimes D_{\alpha_m} \quad \text{where} \quad (\alpha_{j_1},\alpha_{j_2}) < \beta.$$

At least one of these representations must be unitarily equivalent to a direct sum containing D_{α_o}, i.e. there exists $\beta \in A$ such that

$(\alpha_{j_1},\alpha_{j_2}) < \beta$ and

$$(\alpha_1,\ldots,\alpha_{j-1},\beta,\alpha_{j+1},\ldots,\alpha_m) < \alpha_o, \qquad \text{q.e.d.}$$

We must now discuss a second type of symmetries of a Hamiltonian called permutation symmetries. Let π be a permutation of the numbers $1,2,\ldots,N$; define a unitary operator in $L_2(R^{3N})$ by

$$(U(\pi)f)(x) = f(x_{\pi^{-1}(1)},\ldots,x_{\pi^{-1}(N)})$$

for $x \in R^{3N}$ and $f \in L_2(R^{3N})$. Obviously $U(\pi)$ maps $C_o^\infty(R^{3N})$ onto itself. Invariance of an operator under π is defined as for rotations. From the definition of T it follows that T is invariant under π iff $\mu_{\pi(j)} = \mu_j$ for $j=1,\ldots,N$, i.e. iff π permutes only particles of equal mass among themselves. On the other hand, if $\mu_{\pi(j)} = \mu_j$ for all j, the operators $T_{j_1 \ldots j_n}$ transform according to the law

$$U(\pi) T_{\pi(j_1 \ldots j_n)} = T_{j_1 \ldots j_n} U(\pi)$$

where $\pi(j_1 \ldots j_n)$ denotes the ordered image of the subset $\{j_1,\ldots,j_n\}$ under π; if in addition π maps $\{j_1,\ldots,j_n\}$ onto itself, then $T_{j_1 \ldots j_n}$ is invariant under π. For brevity let us say that the operators $T_{j_1 \ldots j_n}$ are covariant under π.

By (6.5') the operators $\tilde{T}_{j_1 \ldots j_n}$ and hence also the operators $\hat{T}_{j_1 \ldots j_n}$ are covariant under π (if $\mu_{\pi(j)} = \mu_j$ for all j). The permutation π is called a permutation symmetry of the Hamiltonian S if the operators $S_{k_1 \ldots k_m}$ and $\hat{S}_{k_1 \ldots k_m}$ are covariant under π; in particular π is a permutation symmetry of S if the operators $T_{j_1 \ldots j_n}$, $\hat{T}_{j_1 \ldots j_n}$, $V_{j_1 \ldots j_n}$ and $W_{j_1 \ldots j_n}$ are all covariant under π. The permutation symmetries of S obviously form a subgroup Π of the symmetric group $\mathcal{S}_N$; we call it the permutation symmetry group of S.

A transposition is a permutation π which interchanges two elements of $\{1, \ldots, N\}$ and leaves all others fixed. The particles i and j are said to be identical, if the transposition which interchages i and j is a permutation symmetry of S. We assume now that Π is generated by transpositions of identical particles. It follows that the system can be decomposed into subsets consisting of identical particles, and consequently Π is a direct product of symmetric groups $\mathcal{S}_{n_j}$ (j=1,...,m) such that $\sum_{j=1}^{m} n_j = N$.

Let E_α ($\alpha \in A$) denote the irreducible unitary (matrix) representations of Π, and let d_α and χ_α be the dimension and the character of E_α respectively. Define

$$(7.7) \qquad P(\Pi, E_\alpha) = \frac{d_\alpha}{h} \sum_{\pi \in \Pi} \chi_\alpha(\pi)^* U(\pi), \quad \alpha \in A$$

where h is the order of Π. It is well known that these operators are orthogonal projections in $L_2(\mathbb{R}^{3N})$ satisfying

$$(7.7') \qquad P(\Pi, E_\alpha) P(\Pi, E_\beta) = \delta_{\alpha\beta} P(\Pi, E_\alpha), \quad \alpha, \beta \in A$$

and

$$(7.7'') \qquad \sum_{\alpha \in A} P(\Pi, E_\alpha) = I.$$

Let Π_o be a subgroup of Π, F_β ($\beta \in B_o$) the irreducible unitary representations of Π_o, and $P(\Pi_o, F_\beta)$ the corresponding orthogonal projections in $L_2(\mathbb{R}^{3N})$. For every $\alpha \in A$ the restriction $E_\alpha|\Pi_o$ is a unitary representation of Π_o and hence unitarily equivalent to a direct sum of irreducible representations:

(7.8) $$E_\alpha|\Pi_o \cong \bigoplus_{\beta \in B_o(\alpha)} n_\beta F_\beta$$

where $B_o(\alpha)$ is a subset of B_o and for every $\beta \in B_o(\alpha)$ the positive integer n_β indicates how many times the representation F_β has to be repeated in the sum.

7.9 Lemma. For a subgroup Π_o of Π and for every $\alpha \in A$ we have

$$P(\Pi,E_\alpha) \leqq \sum_{\beta \in B_o(\alpha)} P(\Pi_o,F_\beta)$$

with $B_o(\alpha)$ defined by (7.8).

Proof. Let $H=L_2(\mathbb{R}^{3N})$. $P(\Pi,E_\alpha)H$ is, by lemma 7.1, an orthogonal sum of E_α-generating subspaces of H. According to (7.8) every E_α-generating subspace is the orthogonal sum of F_β-generating subspaces which, again by lemma 7.1, are contained in $P(\Pi_o,F_\beta)H$. Hence

$$P(\Pi,E_\alpha)H \subset \bigoplus_{\beta \in B_o(\alpha)} P(\Pi_o,F_\beta)H$$

which is equivalent to

$$P(\Pi,E_\alpha) \leqq \sum_{\beta \in B_o(\alpha)} P(\Pi_o,F_\beta).$$

Let Π_1 be a subgroup of Π_o, G_γ $(\gamma \in B_1)$ the irreducible unitary representations of Π_1. As in (7.8) we have

(7.10) $$E_\alpha|\Pi_1 \cong \bigoplus_{\gamma \in B_1(\alpha)} n'_\gamma G_\gamma$$

and

(7.11) $$F_\beta|\Pi_1 \cong \bigoplus_{\gamma \in B_{10}(\beta)} n''_\gamma G_\gamma$$

with certain subsets $B_1(\alpha)$ and $B_{10}(\beta)$ of B_1.

7.12 Lemma. For $B_o(\alpha)$, $B_1(\alpha)$, $B_{10}(\beta)$ defined by (7.8), (7.10) and (7.11) respectively it holds: For every $\gamma \in B_1(\alpha)$ there exists at least one $\beta \in B_o(\alpha)$ such that $\gamma \in B_{10}(\beta)$.

Proof. (7.8) and (7.11) imply

$$E_\alpha|\Pi_1 = (E_\alpha|\Pi_o)|\Pi_1 \cong \bigoplus_{\beta\in B_o(\alpha)} n_\beta \bigoplus_{\gamma\in B_{10}(\beta)} n''_\gamma G_\gamma .$$

From this and (7.10) the result follows immediately.

Let $Z=(Z_1,\ldots,Z_k)$ be a decomposition of the system (i.e. of $\{1,\ldots,N\}$) into k subsystems Z_j, and let $\Pi(Z)$ be the subgroup of Π consisting of all $\pi \in \Pi$ such that $\pi(Z_j)=Z_j$ for $j=1,\ldots,k$. Since Π is generated by transpositions of identical particles, $\Pi(Z)$ is generated by transpositions of pairs of identical particles belonging to one of the subsets Z_j. It follows that $\Pi(Z)=\Pi(Z_1)\times\ldots\times\Pi(Z_k)$, where $\Pi(Z_j)$ is the subgroup generated by transpositions of identical particles in Z_j. Write $L_2(\mathbb{R}^{3N})=\hat{\bigotimes}_{j=1}^{k} L_2(\mathbb{R}^{3m_j})$, where m_j is the number of particles in Z_j. Then for $\pi=\pi_1\times\ldots\times\pi_k\in\Pi(Z)$ we have

$$U(\pi) = \bigotimes_{j=1}^{k} U_j(\pi_j),$$

where $U_j(\pi_j)$ is the unitary operator in $L_2(\mathbb{R}^{3m_j})$ corresponding to $\pi_j \in \Pi(Z_j)$. Apply lemma 7.9 to $\Pi_o=\Pi(Z)$: Since every irreducible representation F_β of $\Pi(Z)$ is the tensorproduct $\bigotimes_{j=1}^{k} F_{\beta j}$ of irreducible representations $F_{\beta j}$ of $\Pi(Z_j)$, we have

$$P(\Pi(Z),F_\beta) = \bigotimes_{j=1}^{k} P(\Pi(Z_j),F_{\beta j})$$

and hence the lemma gives

$$P(\Pi,E) \leq \sum_{\beta\in B_o(\alpha)} \bigotimes_{j=1}^{k} P(\Pi(Z_j),F_{\beta j}).$$

Let us write $(F_1,\dots,F_k) \underset{Z}{\succ} E_\alpha$ instead of $F_\beta = \bigotimes_{j=1}^{k} F_j$, $\beta \in B_o(\alpha)$. Then the last inequality can be rewritten in the form

$$P(\Pi,E_\alpha) \leq \sum_{(F_1,\dots,F_k) \underset{Z}{\succ} E_\alpha} \bigotimes_{j=1}^{k} P(\Pi(Z_j),F_j).$$

For the subspaces

$$L_2(\mathbb{R}^{3N},E_\alpha) = P(\Pi,E_\alpha)L_2(\mathbb{R}^{3N})$$

and

$$L_2(\mathbb{R}^{3m_j},F_j) = P(\Pi(Z_j),F_j)L_2(\mathbb{R}^{3m_j})$$

we have the equivalent relation

$$L_2(\mathbb{R}^{3N},E_\alpha) \subset \bigoplus_{(F_1,\dots,F_k) \underset{Z}{\succ} E_\alpha} \hat{\bigotimes}_{j=1}^{k} L_2(\mathbb{R}^{3m_j},F_j). \tag{7.13}$$

8. The spectrum of the Hamiltonian of a free system

A system is called free if there is no interaction between the particles of the system and external forces, i.e. all operators $V_{j_1 \ldots j_n}$ vanish in the representation (6.0) of the Hamiltonian S. It follows from (6.9), (6.10) that for a free system the total Hamiltonian S and the internal Hamiltonian $\hat{S}$ are related by the equation

$$S = \hat{S} + \tilde{T} \tag{8.1}$$

where $\tilde{T}=\tilde{T}_{1\ldots N}$ is the operator of translation energy of the system. Choose coordinates $y_1,\ldots,y_N$ such that $y_1,\ldots,y_{N-1}$ are internal coordinates for the entire system, and let U be the corresponding unitary operator in $L_2(R^{3N})$ defined by (6.4). Then by definition (6.5) of $\tilde{T}$ the operator $\tilde{T}'=U\tilde{T}U^*$ acts only on y_N and is generated by $B=-(2\mu)^{-1}\Delta$ where Δ is the Laplacean in $L_2(R^3)$ and $\mu=\sum_{j=1}^{N}\mu_j$ is the total mass ot the system. By theorem 6.11 the operator $\hat{S}'=U\hat{S}U^*$ acts only on the internal coordinates and is generated by some operator A in $L_2(R^{3N-3})$. Writing

$$L_2(R^{3N}) = L_2(R^{3N-3})\hat{\otimes}L_2(R^3)$$

and $S'=USU^*$ we get

$$S' = A\otimes I_2 + I_1\otimes B \tag{8.2}$$

where I_1 and I_2 denote the unit operators in $L_2(R^{3N-3})$ and in $L_2(R^3)$ respectively. We derive from (8.2) that S' has the following symmetries:

(8.3) (i) The translations $t_z:(x_1,\dots,x_N) \mapsto (x_1+z,\dots,x_N+z)$, $z \in \mathbb{R}^3$; they form a group isomorphic to $\mathbb{R}^3$. S' is invariant under t_z, since t_z acts according to $(y_1,\dots,y_N) \mapsto (y_1,\dots,y_{N-1},y_N+z)$ in the y-coordinates, and since I_2 and B are translation-invariant.

(ii) The orthogonal transformations $\gamma \in \mathfrak{O}(3)$ acting on the center of mass coordinate only, i.e.

$$(y_1,\dots,y_N) \mapsto (y_1,\dots,y_{N-1},\gamma y_N).$$

Again S' is invariant under these transformations because I_2 and B are invariant.

(iii) The _internal rotations_ $(y_1,\dots,y_N) \mapsto (\gamma y_1,\dots,\gamma y_{N-1},y_N)$ where $\gamma \in \Gamma$, the rotation symmetry group of S defined in section 7, i.e. the subgroup of all $\gamma \in \mathfrak{O}(3)$ under which all operators $W_{j_1\dots j_n}$ are invariant.

(iv) The permutation symmetries π of S. Obviously the center of mass coordinate y_N is invariant under a permutation; hence π acts on the internal coordinates only.

In this section we disregard the translations and we consider the group $\Gamma\times\mathfrak{O}(3)\times\Pi$ generated by the symmetries (iii), (ii) and (iv). The irreducible representations of this group are of the form $R=D_1\otimes D_2\otimes E$ where D_1,D_2,E are irreducible representations of $\Gamma,\mathfrak{O}(3)$ and Π respectively. The corresponding projection P_R in $L_2(\mathbb{R}^{3N})$ with respect to the y-coordinates has the form

$$(8.4) \qquad P_R = P_{D_1}P_E \otimes P_{D_2}$$

where P_{D_1},P_E are the projections in $L_2(\mathbb{R}^{3N-3})$ corresponding to D_1,E and where P_{D_2} in $L_2(\mathbb{R}^3)$ corresponds to D_2.

8.5 Theorem. The projections P_{D_1} and P_E commute. The subspaces

$$L_2(\mathbb{R}^{3N-3},D_1E)=P_{D_1}P_EL_2(\mathbb{R}^{3N-3}),\ L_2(\mathbb{R}^3,D_2)=P_{D_2}L_2(\mathbb{R}^3)$$

$$\text{and} \quad L_2(\mathbb{R}^{3N}, R) = P_R L_2(\mathbb{R}^{3N})$$

reduce the operators A, B and S' respectively. For the reduced operators $A_{D_1 E}$, B_{D_2} and S'_R we have

$$(8.6) \qquad S'_R = A_{D_1 E} \otimes I_2 + I_1 \otimes B_{D_2}$$

<u>Proof.</u> P_E is a finite linear combination of unitary operators $U(\pi)$ $(\pi \in \Pi)$ satisfying $U(\pi)A = AU(\pi)$; hence P_E satisfies $P_E A \subset AP_E$, i.e. $P_E D(A) \subset D(A)$ and $P_E Au = AP_E u$ for $u \in D(A)$, and consequently the subspace $P_E L_2(\mathbb{R}^{3N-3})$ reduces A. P_{D_1} commutes with P_E, and $P_{D_1} L_2(\mathbb{R}^{3N-3})$ reduces A. This is obvious if Γ is a finite subgroup of $\mathfrak{O}(3)$, since in this case P_{D_1} is a finite linear combination of unitary operators $U(\gamma)$ $(\gamma \in \Gamma)$ satisfying $U(\gamma)P_E = P_E U(\gamma)$ and $U(\gamma)A = AU(\gamma)$. If Γ is infinite we have

$$P_{D_1} = d \int_\Gamma \chi(\gamma)^* U(\gamma)\, d\mu(\gamma)$$

where d and χ are the dimension and the character of D_1 respectively and where μ is the invariant probability measure on Γ. It follows that P_{D_1} commutes with P_E. For $f \in C_o^\infty(\mathbb{R}^{3N-3})$ there is a compact subset K of $\mathbb{R}^{3N-3}$ such that $\mathrm{supp}(U(\gamma)f) \subset K$ for all $\gamma \in \mathfrak{O}(3)$; hence $P_{D_1} f$ has compact support. Also $U(\gamma)f \in C_o^\infty(\mathbb{R}^{3N-3})$ and $\mathrm{grad}_j[U(\gamma)f] = U(\gamma)[\mathrm{grad}_j f \cdot \gamma^{-1}]$ for $j = 1, 2, \ldots, N-1$. It follows that

$$P_{D_1} f = d \int_\Gamma \chi(\gamma)^* U(\gamma) f\, d\mu(\gamma)$$

is infinitely differentiable and the derivatives can be computed by differentiation of the integrand, hence $P_{D_1} f \in C_o^\infty(\mathbb{R}^{3N-3}) = D(A)$. Finally $P_{D_1} f$ is the limit in $L_2(\mathbb{R}^{3N-3})$ of finite linear combinations of elements $U(\gamma)f$ $(\gamma \in \Gamma)$. Since A is closable this implies $P_{D_1} Af = AP_{D_1} f$, i.e. $P_{D_1}L_2(\mathbb{R}^{3N-3})$ reduces A; hence $P_{D_1}P_E L_2(\mathbb{R}^{3N-3})$ also reduces A. The statement concerning P_{D_2} and B is proved exactly as for P_{D_1} and A. P_R defined by (8.4) obviously is a projection in $L_2(\mathbb{R}^{3N})$ satisfying $P_R S' \subset S' P_R$ according to (8.2). Hence $P_R L_2(\mathbb{R}^{3N})$ reduces S'. Finally (8.6) follows from (8.2) and (8.4).

8.7 Theorem. Let D_1, D_2 and E be such that $P_R \neq 0$. The operators S'_R, A_{D_1E} and B_{D_2} are essentially selfadjoint. The spectrum $\sigma(S'_R)$ of S' (i.e. of the selfadjoint closure $\overline{S'_R}$) is the closure of the set

$$\{\lambda : \lambda = \lambda_1 + \lambda_2, \lambda_1 \in \sigma(A_{D_1E}), \lambda_2 \in \sigma(B_{D_2})\}.$$

Proof. By assumption (6.12) and lemma 6.6 the operators S, $\hat{S}$ and $\tilde{T}$ are essentially selfadjoint; hence the same is true for S', $\hat{S}'$ and $\tilde{T}'$. By theorem 5.2 the operators A and B (which generate $\hat{S}'$ and $\tilde{T}'$ respectively) are also essentially selfadjoint. Hence the reduced operators S'_R, A_{D_1E} and B_{D_2} are essentially selfadjoint (remember that P_R and hence $P_{D_1}P_E$ and P_{D_2} do not vanish). The assertion concerning the spectrum of S'_R is a consequence of (8.6) and of a theorem on separable operators (see K.Maurin [20],II.6.th.8).

The next theorem concerns $\sigma(B_{D_2})$; we prove it in more generality than is needed here because this can be done without extra work.

8.8 Theorem. Let $T = -\Delta$ in $L_2(\mathbb{R}^m)$, $D(T) = C_o^\infty(\mathbb{R}^m)$, Γ a compact subgroup of $\mathfrak{O}(m)$, D_α an irreducible representation of Γ such that the corresponding projection P_α does not vanish, T_α the reduced operator (with domain $P_\alpha C_o^\infty(\mathbb{R}^m)$). Then

$$\sigma(T_\alpha) = [0, \infty[;$$

for every $\lambda \geq 0$, $\varepsilon > 0$ and $r > 0$ there exists a D_α-generating subspace N of D(T) such that

$$u(x)=0 \text{ for } |x|\leq r \quad\text{and}\quad \|Tu-\lambda u\|\leq \varepsilon\|u\| \quad\text{for all } u \in N.$$

Proof. The last statement is proved first: Let Ω denote the unit sphere in $\mathbb{R}^m$, and let N_ℓ be the set of spherical harmonics of order ℓ on Ω; N_ℓ is a finite-dimensional subspace of $L_2(\Omega)$ and

$$\bigoplus_{\ell=0}^{\infty} N_\ell = L_2(\Omega).$$

Let d_ℓ be the dimension of N_ℓ and let $Y_{\ell j}$, $j=1,\ldots,d_\ell$ be an orthonormal basis. For every $\gamma \in \mathfrak{G}(m)$ the function $\xi \mapsto Y_{\ell j}(\gamma^{-1}\xi)$ is a spherical harmonic of order ℓ; hence there is a unitary matrix $D^{(\ell)}(\gamma) = (D^{(\ell)}_{ij}(\gamma))$ such that

$$Y_{\ell j}(\gamma^{-1}\xi) = \sum_{i=1}^{d_\ell} D^{(\ell)}_{ij}(\gamma) Y_{\ell i}(\xi), \qquad \xi \in \Omega,\ j=1,\ldots,d_\ell$$

and it is easy to see that $\gamma \mapsto D^{(\ell)}(\gamma)$ is a unitary representation of Γ. Since the representation $D^{(\ell)}$ is unitarily equivalent to a direct sum of irreducible representations, the subspace N_ℓ is the orthogonal sum of D_β-generating subspaces of $L_2(\Omega)$ for some irreducible representations D_β of Γ. But $\bigoplus_{\ell=0}^{\infty} N_\ell = L_2(\Omega)$ and $P_\alpha \neq 0$; therefore there is at least one ℓ such that N_ℓ contains a D_α-generating subspace $N(\alpha,\ell)$. For $f \in L_2(0,\infty)$ with $\|f\|=1$ let $N(\alpha,\ell,f)$ be the set of functions u on $\mathbb{R}^m$ of the form

$$u(x) = |x|^{\frac{1-m}{2}} f(|x|) Y\left(\frac{x}{|x|}\right), \qquad x \in \mathbb{R}^m$$

where $Y \in N(\alpha,\ell)$. Then $N(\alpha,\ell,f)$ is a D_α-generating subspace of $L_2(\mathbb{R}^m)$. For $f \in C_o^\infty(0,\infty)$ we have $N(\alpha,\ell,f) \subset D(T)$ and

$$Tu(x) = |x|^{\frac{1-m}{2}} L_\ell f(|x|) Y\left(\frac{x}{|x|}\right)$$

where

$$L_\ell f(s) = -f''(s) + \left(\ell + \frac{m-1}{2}\right)\left(\ell + \frac{m-3}{2}\right) s^{-2} f(s).$$

Now put

$$f_n(s) = n^{-1/2} e^{i\sqrt{\lambda}s} \varphi(\tfrac{s}{n}), \qquad n=1,2,\ldots$$

where $\lambda \geq 0$, $\varphi \in C_o^\infty(0,\infty)$ and $\|\varphi\|=1$. Then

$$\|f_n\|=1 \quad \text{and} \quad \|L_\ell f_n - \lambda f_n\| \to 0 \quad \text{for } n \to \infty .$$

Given $\varepsilon > 0$ and $r > 0$ we can find $n \in \mathbb{N}$ such that

$$f_n(s)=0 \quad \text{for } s \leq r \quad \text{and} \quad \|L_\ell f_n - \lambda f_n\| \leq \varepsilon.$$

It follows that $N=N(\alpha,\ell,f_n)$ is a D_α-generating subspace of $D(T)$ such that

$$u(x)=0 \quad \text{for } |x| \leq r \quad \text{and} \quad \|Tu-\lambda u\| \leq \varepsilon\|u\| \quad \text{for all } u \in N.$$

By lemma 7.1, N is contained in $P_\alpha L_2(\mathbb{R}^m)$ and hence $N \subset D(T_\alpha)$. By theorem 1.1 part (3') we have $\lambda \in \sigma(T_\alpha)$; this is true for every $\lambda \geq 0$, hence $[0,\infty[\subset \sigma(T_\alpha)$. On the other hand $\sigma(T)=[0,\infty[$ implies $\sigma(T_\alpha) \subset [0,\infty[$, hence $\sigma(T_\alpha)=[0,\infty[$.

<u>8.8' Corollary.</u> Let $P_\pm$ be the orthogonal projections in $L_2(\mathbb{R}^m)$ defined by

$$(P_\omega f)(x) = \tfrac{1}{2} f(x) + \tfrac{\omega}{2} f(-x) \qquad \text{for } \omega=\pm 1.$$

If $P_\omega P_\alpha \neq 0$, then the subspace N in theorem 8.8 can be chosen such that $P_\omega N=N$.

The proof is an easy consequence of the fact that the subspaces $N(\alpha,\ell,f)$ defined in the proof of theorem 8.8 satisfy $P_\omega N(\alpha,\ell,f) = N(\alpha,\ell,f)$ if $\omega=(-1)^\ell$ and $P_\omega N(\alpha,\ell,f)=\{0\}$ if $\omega \neq (-1)^\ell$.

Let D and E be irreducible representations of the rotation symmetry group Γ and of the permutation symmetry group Π of S respectively. Denote by P_D and P_E the corresponding projections in $L_2(\mathbb{R}^{3N})$. Then $L_2(\mathbb{R}^{3N},DE)=P_D P_E L_2(\mathbb{R}^{3N})$ is a reducing subspace of S; we want to find the spectrum of the reduced operator S_{DE}. The relation $(\alpha_1,\alpha_2)<\alpha$ was defined in section 7 for irreducible representations $D_{\alpha_1}, D_{\alpha_2}, D_\alpha$ of Γ (see (7.2) and (7.2')); write $(D_{\alpha_1}, D_{\alpha_2})<D_\alpha$ for this relation. Using the notation of theorem 8.7 we define the number

$$\eta_{DE} = \inf\{\lambda : \lambda \in \sigma(A_{D_1E}),\ \text{there exists } D_2 \text{ such that} \quad (8.9)$$
$$(D_1,D_2) < D \ \text{and}\ P_{D_1}P_EP_{D_2} \neq 0\}.$$

The infimum is finite because $\sigma(A_{D_1E}) \subset \sigma(A)=\sigma(S)$ and because S is bounded below (see (6.12)).

<u>8.10 Theorem.</u> Let S be the Hamiltonian of a free system, and let D and E be irreducible representations of the rotation symmetry group Γ and of the permutation symmetry group Π of S respectively, such that the corresponding projection P_DP_E does not vanish. For the reduced operator S_{DE} we have

$$\sigma(S_{DE}) = [\eta_{DE}, \infty[$$

where η_{DE} is given by (8.9). For every $\lambda \geq \eta_{DE}$ and $\varepsilon > 0$ there exists a number $\rho > 0$ such that for every $r > 0$ there is a $D\otimes E$-generating subspace N of D(S) such that for all $u \in N$ we have

$$\|Su-\lambda u\| \leq \varepsilon\|u\|,$$

and

$$u(x)=0 \quad \text{for} \quad \Big(\sum_{i,j=1}^{N} |x_i-x_j|^2\Big)^{1/2} \geq \rho \quad \text{and for} \quad \mu^{-1}\Big|\sum_{j=1}^{N}\mu_j x_j\Big| \leq r.$$

<u>Proof</u>. a) Let (D_1^i, D_2^i) be all pairs of irreducible representations of Γ such that $(D_1^i,D_2^i)<D$ and $P_{D_1^i}P_EP_{D_2^i} \neq 0$. For $R_i=D_1^i \otimes D_2^i \otimes E$ use the representation (8.4) of the corresponding projection P_{R_i} and define the reducing subspace $L_2(\mathbb{R}^{3N},R_i)=P_{R_i}L_2(\mathbb{R}^{3N})$ of S'. Apply theorem 8.7 to the reduced operators S'_{R_i}: Since $\sigma(B_{D_2^i})=[0,\infty[$ by theorem 8.8, we get

$$\sigma(S'_{R_i}) = [\mu_{D_1^iE}, \infty[\quad \text{where } \mu_{D_1^iE} = \min \sigma(A_{D_1^iE}) \geq \eta_{DE}.$$

Now let $L_2(\mathbb{R}^{3N}, DE) = P_D P_E L_2(\mathbb{R}^{3N})$; by lemma 7.5 we have

$$L_2(\mathbb{R}^{3N}, DE) \subset \underset{i}{\oplus} L_2(\mathbb{R}^{3N}, R_i)$$

and hence $\sigma(S_{DE}) \subset \bigcup_i \sigma(S'_{R_i}) \subset [\eta_{DE}, \infty[$.

b) Let $\lambda > \eta_{DE}$, then according to (8.9) there exists a pair (D_1, D_2) such that $(D_1, D_2) < D$, $P_{D_1} P_E P_{D_2} \neq 0$ and $\lambda \geq \mu_{D_1E} = \min \sigma(A_{D_1E})$. Let d_1, d_2 and d_3 be the dimension of D_1, D_2 and E respectively. Let $\varepsilon > 0$ and $r > 0$ be given and put $\varepsilon' = \frac{1}{2}\varepsilon(d_1 d_2 d_3)^{-1/2}$; by theorem 8.8 there exists a D_2-generating subspace N_2 of $D(B) = C_o^\infty(\mathbb{R}^3)$ such that $w(x) = 0$ for $|x| \leq r$ and

$$\|Bw - (\lambda - \mu_{D_1E})w\| \leq \varepsilon' \|w\| \qquad \text{for all } w \in N_2.$$

Now $\mu_{D_1E} \in \sigma(A_{D_1E})$; hence, by theorem 8.11 below, there exists a $D_1 \otimes E$-generating subspace N_1 of $D(A) = C_o^\infty(\mathbb{R}^{3N-3})$ such that

$$\|Av - \mu_{D_1E} v\| \leq \varepsilon' \|v\| \qquad \text{for all } v \in N_1.$$

By (8.2) it follows that $\|S'u - \lambda u\| \leq 2\varepsilon' \|u\|$ for all $u = v \otimes w$, where $v \in N_1$ and $w \in N_2$. Choose orthonormal bases $v_1, \dots v_p$ $(p = d_1 d_3)$ of N_1 and $w_1, \dots, w_q$ $(q = d_2)$ of N_2. Every $u \in N_1 \otimes N_2$ is of the form

$$u = \sum_{j=1}^{p} \sum_{k=1}^{q} c_{jk} v_j \otimes w_k$$

and hence

$$\|S'u - \lambda u\| \leq 2\varepsilon' \sum_{j=1}^{p} \sum_{k=1}^{q} |c_{jk}|$$

$$\leq 2\varepsilon' (pq)^{1/2} \{ \sum_{j=1}^{p} \sum_{k=1}^{q} |c_{jk}|^2 \}^{1/2} = \varepsilon \|u\|.$$

By construction we have $u(y)=0$ for $|y_N| \leq r$ for all $u \in N_1 \otimes N_2$, and there is a number $\rho' > 0$ such that

$$u(y)=0 \quad \text{for} \quad \left(\sum_{j=1}^{N-1} |y_j|^2\right)^{1/2} \geq \rho' \quad \text{for all } u \in N_1 \otimes N_2.$$

By definition of the relation $(D_1, D_2) \prec D$ and by lemma 7.3, $N_1 \otimes N_2$ contains a $D\otimes E$-generating subspace M. Transforming back to the original variable x we get a $D\otimes E$-generating subspace N of D(S) such that for all $u \in N$ we have

$$\|Su-\lambda u\| \leq \epsilon \|u\|,$$

$$u(x)=0 \quad \text{for} \quad \mu^{-1} \left| \sum_{j=1}^{N} \mu_j x_j \right| \leq r,$$

and by lemma 6.8,

$$u(x)=0 \quad \text{for} \quad \left(\sum_{i,j=1}^{N} |x_i - x_j|^2 \right)^{1/2} \geq \rho = \frac{\rho'}{c_1}.$$

Let now $\lambda = \eta_{DE}$; then for $\epsilon > 0$ there exists a $\rho > 0$ such that for every $r > 0$ there is a $D\otimes E$-generating subspace N of D(S) such that for all $u \in N$ we have

$$\|Su-(\lambda+\tfrac{\epsilon}{2})u\| \leq \tfrac{\epsilon}{2}\|u\|$$

and

$$u(x)=0 \quad \text{for} \quad \left(\sum_{i,j=1}^{N} |x_i - x_j|^2 \right)^{1/2} \geq \rho \quad \text{and for} \quad \mu^{-1} \left| \sum_{j=1}^{N} \mu_j x_j \right| \leq r.$$

This implies $\|Su-\lambda u\| \leq \epsilon \|u\|$ for all $u \in N$.
Finally it follows from theorem 1.1 part (3') that $\lambda \in \sigma(S_{DE})$, i.e. $\sigma(S_{DE}) \supset [\eta_{DE}, \infty[$, and hence $\sigma(S_{DE}) = [\eta_{DE}, \infty[$.

The following theorem has been used in the proof of theorem 8.10; we state and prove it in more generality than needed, because of its intrinsic usefulness. For the application in the proof of theorem 8.10 one has to replace Γ by $\Gamma \times \Pi$ and D by $D_1 \otimes E$.

8.11 Theorem. Let A be an essentially selfadjoint operator in $L_2(\mathbb{R}^m)$ with $D(A)=C_o^\infty(\mathbb{R}^m)$, and let Γ denote a compact group of transformations of $\mathbb{R}^m$ such that $(U(\gamma)f)(x)=f(\gamma^{-1}x)$ defines a continuous representation $\gamma \mapsto U(\gamma)$ of Γ by unitary operators $U(\gamma)$ in $L_2(\mathbb{R}^m)$ satisfying $U(\gamma)A = AU(\gamma)$ for all $\gamma \in \Gamma$. Then for every irreducible unitary (matrix) representation D of Γ the corresponding subspace $P_D L_2(\mathbb{R}^m)$ reduces A. For every $\lambda \in \sigma(A_D)$, where A_D is the reduced operator, and for every $\varepsilon > 0$ there exists a D-generating subspace N of D(A) such that

$$\|Au-\lambda u\| \leq \varepsilon\|u\| \quad \text{for all } u \in N.$$

Proof. a) Let d be the dimension of D and let $D:\gamma \mapsto (D_{jk}(\gamma))$ where $j,k=1,\dots,d$. Let μ denote the invariant probability measure on Γ and define continuous operators P_{jk} in $L_2(\mathbb{R}^m)$ by

$$P_{jk} = d\int_\Gamma D_{jk}(\gamma)^* U(\gamma)\,d\mu(\gamma), \qquad j,k=1,\dots,d.$$

It follows from the properties of $D_{jk}(\cdot)$ and $U(\cdot)$ that

$$U(\gamma)P_{jk} = \sum_{r=1}^{d} D_{rj}(\gamma)P_{rk} \tag{8.12}$$

for all $\gamma \in \Gamma$ and for $j,k=1,\dots,d$. Furthermore, using (7.0") with α,β omitted, one verifies the relations

$$P_{jk}P_{pq} = \delta_{kp}P_{jq} \tag{8.13}$$

$$P^*_{jk}P_{pq} = \delta_{jp}P_{kq} \tag{8.14}$$

$$P_{jk}P^*_{pq} = \delta_{kq}P_{jp} \tag{8.15}$$

for all $j,k,p,q=1,\dots,d$. (8.13) and (8.14) imply

$$P_{jj}P_{pp} = \delta_{jp}P_{jj} \quad \text{and} \quad P^*_{jj}P_{jj} = P_{jj};$$

hence P_{jj} are orthogonal projections onto mutually orthogonal subspaces $P_{jj}L_2(\mathbb{R}^m)$. The sum

$$P_D = \sum_{j=1}^{d} P_{jj} = d\int_\Gamma \chi(\gamma)^* U(\gamma)\,d\mu(\gamma)$$

is the orthogonal projection corresponding to the representation D. From (8.14) and (8.15) we get

$$P^*_{jk}P_{jk} = P_{kk} \quad \text{and} \quad P_{jk}P^*_{jk} = P_{jj}$$

which shows that P_{jk} is partially isometric with initial set $P_{kk}L_2(\mathbb{R}^m)$ and final set $P_{jj}L_2(\mathbb{R}^m)$ (compare T. Kato [15],V.2.2).

Let $f \neq 0$ be an element of $P_DL_2(\mathbb{R}^m)$; then there exists $q\in\{1,\dots,d\}$ such that $P_{qq}f \neq 0$. Define $g=\|P_{qq}f\|^{-1}P_{qq}f$ and $u_j=P_{jq}g$ for $j=1,\dots,d$. It follows that $u_j \in P_{jj}L_2(\mathbb{R}^m)$ and $\|u_j\|=1$, hence $\langle u_j|u_k\rangle=\delta_{jk}$, and (8.12) implies

$$U(\gamma)u_j = \sum_{r=1}^{d} D_{rj}(\gamma)u_r \quad \text{for } \gamma \in \Gamma \quad \text{and } j=1,\dots,d.$$

This shows that $u_1,\dots,u_d$ is the orthonormal basis for a D-generating subspace of $L_2(\mathbb{R}^m)$.

b) The operators P_{jk} have the further property

$$(8.16) \qquad P_{jk}A \subset AP_{jk}, \quad j,k=1,\dots,d;$$

in particular the subspaces $P_{jj}L_2(\mathbb{R}^m)$ reduce A, and hence $P_DL_2(\mathbb{R}^m)$ reduces A. Comparing with the proof of theorem 8.5 it suffices to show that for every $f \in C_o^\infty(\mathbb{R}^m)$ the set $K=\bigcup_{\gamma\in\Gamma}$ supp $(U(\gamma)f)$ is compact.

Let $\mathbb{V}$ be an open covering of K; by assumption supp $(U(\gamma)f)=K_\gamma$ is compact, hence for every $\gamma\in\Gamma$ there exists a finite subset $\mathbb{V}_\gamma$ of $\mathbb{V}$ such that $K_\gamma \subset \bigcup\{V: V \in \mathbb{V}_\gamma\}$. By definition of $U(\gamma)$ the set K_γ is the image of supp (f) under γ; hence for every $\gamma \in \Gamma$ there exists an open neighbourhood Σ_γ such that $K_\delta \subset \bigcup\{V: V \in \mathbb{V}_\gamma\}$ for all $\delta \in \Sigma_\gamma$. $\{\Sigma_\gamma: \gamma \in \Gamma\}$ being an open covering of Γ contains a finite covering $\{\Sigma_{\gamma_1},\dots,\Sigma_{\gamma_n}\}$.

It follows that $\bigcup_{i=1}^{n} \mathbb{W}_{\gamma_i}$ is a finite open covering of K contained in $\mathbb{V}$; hence K is compact.

c) Let $\lambda \in \sigma(A_D)$ and $\varepsilon > 0$ be given. Then by theorem 1.1 there exists $f \in D(A_D)=P_D D(A)$ such that

$$\|f\|=1 \quad \text{and} \quad \|Af-\lambda f\| \leq d^{-1/2}\varepsilon.$$

Now $f=P_D f=\sum_{j=1}^{d} P_{jj}f$ implies $1=\sum_{j=1}^{d}\|P_{jj}f\|^2$; hence there exists $q\in\{1,\ldots,d\}$ such that $\|P_{qq}f\|^2 \geq d^{-1}$. Define $g=\|P_{qq}f\|^{-1}P_{qq}f$; then $\|g\|=1$, and (8.16) gives

$$\|Ag-\lambda g\|=\|P_{qq}(Af-\lambda f)\|\,\|P_{qq}f\|^{-1} \leq d^{-1/2}\varepsilon d^{1/2}=\varepsilon.$$

By part a) of the proof the elements $u_j=P_{jq}g$ form an orthonormal basis of a D-generating subspace N of $L_2(\mathbb{R}^m)$. By (8.16) we have $N \subset D(A)$ and

$$\|Au_j-\lambda u_j\|=\|P_{jq}(Ag-\lambda g)\| \leq \varepsilon \quad \text{for } j=1,\ldots,d.$$

It follows that $\|Au-\lambda u\| \leq \varepsilon\|u\|$ for all $u \in N$, q.e.d.

9. A lower bound of the essential spectrum

In this section we present a rather technical theorem (th.9.4) on the essential spectrum of a Hamiltonian which, however, is the decisive step towards the final results in section 10 and 11.

Let $p_{hklr'}$ denote a family of infinitely differentiable real functions on $\mathbb{R}^m$ for

$$h \in \{1,\dots,H\}, \quad k \in \{1,\dots,k(h)\},$$

$$\ell \in \{1,\dots,\ell(h,k)\}, \quad r' \in]0,\infty[$$

such that $p_{hk\ell r'}$ has derivatives of first and second order which are bounded uniformly on $\mathbb{R}^m$ and for all values of the parameters h,k,ℓ,r'. Define open subsets of $\mathbb{R}^m$ by

(9.0) $\qquad \Omega(h,k,r',r)=\{x \in \mathbb{R}^m : p_{hk\ell r'}(x)>r, \quad \ell=1,\dots,\ell(h,k)\}$

and assume the following:

(9.1) There exists $\alpha > 0$ such that

(i) $d(\Omega(h,k,r',s),\Omega(h,k',r',t)) > \alpha^{-1}(s+t-r')$ for all $h,k\neq k'$, $s > 0$, $t > 0$, $s+t > r' > 0$

and

(ii) $d(\Omega(h,k,r',s),\complement\Omega(h,k,r',t)) > \alpha^{-1}(s-t)$ for all h,k, $s> t >0$ and $r' > 0$.

(9.2) $\qquad \complement(\bigcup_{h,k}\Omega(h,k,r',r)) \subset K(r')$ for $0 < r < r'$, where $K(r')$ is a compact subset of $\mathbb{R}^m$ depending on r' only.

By definition we have $\Omega(h,k,r',s) \subset \Omega(h,k,r',r)$ for $r \leq s$; hence we can define

$$\Psi(h,k,r',r,s) = \Omega(h,k,r',r) \setminus \Omega(h,k,r',s)$$

for $r \leq s$.

9.3 Lemma. Given $C > 0$, $\varepsilon \in]0,1]$ and $\rho > 0$ there exists $r' > \rho+2$ such that for every nonnegative Borel measure B on $\mathbb{R}^m$ satisfying $B(\mathbb{R}^m) \leq C$ there exists a number r such that $r+\alpha\rho+2 < r' \leq 2r$ and

$$\sum_{h,k} B(\Psi(h,k,r',r,r+\alpha\rho+2)) \leq \varepsilon^2.$$

Proof. Choose $J \in \mathbb{N}$ such that $J \geq \varepsilon^{-2}CH$; let

$$r_j = \rho+(J+j)(\alpha\rho+2) \quad \text{for } j=0,1,\ldots,J \quad \text{and}$$

$$r' = r_J+\rho.$$

Then

$$r' > \rho+2, \qquad r_j+\alpha\rho+2 \leq r_J < r' \quad \text{and}$$

$$2r_j \geq 2r_o = 2\rho+2J(\alpha\rho+2) = r' \quad \text{for } j=0,1,\ldots,J-1.$$

We have

$$\Psi(h,k,r',r_j,r_{j+1}) \cap \Psi(h,k',r',r_i,r_{i+1}) = \emptyset$$

for $k=k'$, $j \neq i$ by definition of the sets Ψ; the same holds for $k \neq k'$ and arbitrary i,j because

$$\Omega(h,k,r',r_j) \cap \Omega(h,k',r',r_i) = \emptyset$$

by (9.1.i) since $r_j+r_i \geq r'$ for $i,j=0,\ldots,J-1$. It follows that

$$\sum_{j=0}^{J-1} \sum_{k=1}^{k(h)} B(\Psi(h,k,r',r_j,r_{j+1})) \leq B(\mathbb{R}^m) \leq C$$

and hence

$$\sum_{j=0}^{J-1} \sum_{h,k} B(\Psi(h,k,r',r_j,r_{j+1})) \leq CH.$$

Consequently there exists $j_o \in \{0,\dots,J-1\}$ such that

$$\sum_{h,k} B(\Psi(h,k,r',r_{j_o},r_{j_o+1})) \leq J^{-1}CH \leq \varepsilon^2,$$

i.e. the lemma is proven with $r=r_{j_o}$.

<u>9.4 Theorem.</u> Let $S=T+V$ be an essentially selfadjoint operator in $L_2(\mathbb{R}^m)$ with $D(S)=C_o^\infty(\mathbb{R}^m)$, where $T=-\Delta$ and V is symmetric and T-bounded as well as S-bounded. Assume furthermore:

(i) For every $\varepsilon > 0$ there exists $\rho_\varepsilon > 0$ such that

$$|\langle Vu|v\rangle| \leq \varepsilon\{\|u\|^2+\|Tu\|^2+\|v\|^2+\|Tv\|^2\}$$

for all $u,v \in C_o^\infty(\mathbb{R}^m)$ satisfying $d(\operatorname{supp}(u), \operatorname{supp}(v)) \geq \rho_\varepsilon$.

(ii) For every pair h,k ($h \in \{1,\dots,H\}$, $k \in \{1,\dots,k(h)\}$) there is a symmetric operator V_{hk} with $D(V_{hk}) \supset D(S)$ such that $S_{hk}=S-V_{hk}$ is essentially selfadjoint and such that the following holds: For every $\varepsilon > 0$ there exists $r_\varepsilon > 0$ such that $\|V_{hk}u\| \leq \varepsilon(\|u\|+\|Tu\|)$ for all $u \in C_o^\infty(\mathbb{R}^m)$ satisfying $\operatorname{supp}(u) \subset \Omega(h,k,r',s)$ with numbers s,r' such that $r_\varepsilon < r' \leq 2s$ and $s \leq r'-1$.

Assertion: If $\eta_{hk}=\min \sigma(S_{hk})$ and $\eta = \min\{\eta_{hk}: h=1,\dots,H;\ k=1,\dots,k(h)\}$ then $\sigma_e(S) \subset [\eta,\infty[$.

Proof. a) Let $\lambda \in \sigma_e(S)$; then by theorem 1.2 (4') there exists a sequence (u_n) in $D(S)$ such that $\|u_n\|=1$, $u_n \rightharpoonup 0$ and $Su_n-\lambda u_n \to 0$. This implies $\langle u_n|Su_n\rangle \to \lambda$; hence we have to show that $\lim_{n\to\infty} \langle u_n|Su_n\rangle \geq \eta$. The sequences (u_n) and (Su_n) are bounded, and $T=S-V$ is S-bounded by assumption; hence (Tu_n) is also bounded, and there exists a constant C such that

$$\|u_n+Tu_n\|^2 = \int\{|u_n|^2+2|\operatorname{grad} u_n|^2+|Tu_n|^2\}dx \leq C$$

for all n. For every $u \in C_o^\infty(\mathbb{R}^m)$ define a nonnegative Borel measure B_u on $\mathbb{R}^m$ by

$$B_u(\Omega) = \int_\Omega \{|u|^2+2|\text{grad } u|^2+|Tu|^2\}dx.$$

It follows that $B_{u_n}(\mathbb{R}^m) \leq C$ for all n.

b) We now define a decomposition of the functions u_n; for convenience of notation we drop the index n throughout this part of the proof. Given $\varepsilon \in]0,1]$, we choose ρ_ε and r_ε according to assumptions (i) and (ii) and we define $\rho=\max\{\rho_\varepsilon, r_\varepsilon\}$. According to lemma 9.3 we choose r' (independent of u) and then r depending on $B=B_u$. Let $\varphi \in C^\infty(\mathbb{R})$ satisfy $0 \leq \varphi(t) \leq 1$ for all $t \in \mathbb{R}$, $\varphi(t)=0$ for $t \leq -1$, and $\varphi(t)=1$ for $t \geq 0$. Define

$$\Phi_{hks}(x) = \prod_{\ell=1}^{\ell(h,k)} \varphi(p_{hk\ell r'}(x)-r-s-1) \tag{9.5}$$

for $x \in \mathbb{R}^m$, $s \geq 0$, $h=1,\ldots,H$ and $k=1,\ldots,k(h)$. It follows that

$$\Phi_{hks} \in C^\infty(\mathbb{R}^m), \quad 0 \leq \Phi_{hks}(x) \leq 1 \quad \text{for all } x \in \mathbb{R}^m$$

and (according to (9.0))

$$\Phi_{hks}(x) = \begin{cases} 0 & \text{for } x \in \complement\Omega(h,k,r',r+s) \\ 1 & \text{for } x \in \Omega(h,k,r',r+s+1), \end{cases} \tag{9.6}$$

hence we have

$$\text{supp}(\Phi_{hks}) \subset \overline{\Omega(h,k,r',r+s)} \tag{9.7}$$

and (9.1.i) implies

$$d(\text{supp}(\Phi_{hks}), \text{supp}(\Phi_{hjt})) \geq \alpha^{-1}(s+t) \quad \text{for } j\neq k \tag{9.8}$$

(note that $2r \geq r'$ by lemma 9.3).
Define functions v_h, $w_{hk} \in C_o^\infty(\mathbb{R}^m)$ by $v_o=u$ and

$$v_h=v_{h-1}-\sum_{k=1}^{k(h)} w_{hk}, \quad w_{hk}=v_{h-1}\Phi_{hks_o} \tag{9.9}$$

for $h=1,\ldots,H$, where $s_o=\alpha\rho+1$. For $x \in \Omega(h,k,r',r+\alpha\rho+2)$ we have

$\Phi_{hjs_o}(x)=\delta_{jk}$ and therefore $v_h(x)=v_{h-1}(x)[1-\Phi_{hks_o}(x)]=0$, hence

$$(9.9') \qquad \mathrm{supp}(v_h) \subset \mathrm{supp}(v_{h-1}) \cap \complement[\bigcup_{k=1}^{k(h)} \Omega(h,k,r',r+\alpha\rho+2)]$$

which implies

$$(9.10) \qquad \mathrm{supp}(v_H) \subset \complement(\bigcup_{h,k} \Omega(h,k,r',r+\alpha\rho+2)) \subset K(r')$$

by (9.2), where $K(r')$ is compact (note that $r+\alpha\rho+2< r'$ by lemma 9.3). We need some estimates: By assumption there exists a bound γ_o for the derivatives of first and second order of $p_{hk\ell r'}$ uniformly for all h,k,ℓ,r'; hence there is a positive number γ, depending only on γ_o and on the function φ, such that

$$(9.11) \qquad |\mathrm{grad}\, \Phi_{hks}(x)| \leq \gamma, \quad |\Delta\Phi_{hks}(x)| \leq \gamma, \qquad x \in \mathbb{R}^m$$

for all h,k and s. It follows the existence of a number γ', depending on γ and H only, such that

$$(9.11') \qquad B_{v_h}(\Omega) \leq \gamma' B_u(\Omega), \quad B_{w_{hk}}(\Omega) \leq \gamma' B_u(\Omega)$$

for all Borel sets $\Omega \subset \mathbb{R}^m$ and for all h and k. For $j \neq k$ the functions w_{hj} and w_{hk} have disjoint supports, hence we get $\langle w_{hj}|w_{hk}\rangle=0$ and

$$\|v_{h-1}\|^2=\|v_h\|^2+\sum_{k=1}^{k(h)} \|w_{hk}\|^2+2\,\mathrm{Re}\sum_{k=1}^{k(h)} \langle v_h|w_{hk}\rangle.$$

Now

$$\mathrm{supp}(v_h) \cap \mathrm{supp}(w_{hk}) \subset \overline{\Omega(h,k,r',r+\alpha\rho+1)} \setminus \Omega(h,k,r',r+\alpha\rho+2)$$

$$\subset \overline{\Psi(h,k,r',r,r+\alpha\rho+2)}.$$

Denote this set by Ψ_{hk}; then by definition of B_u we get

$$|2\,\mathrm{Re}\sum_{k=1}^{k(h)} \langle v_h|w_{hk}\rangle| \leq \sum_k \int_{\Psi_{hk}} \{|v_h|^2+|w_{hk}|^2\}dx$$

$$\leq \sum_k [B_{v_h}(\Psi_{hk})+B_{w_{hk}}(\Psi_{hk})] \leq 2\gamma'\sum_k B_u(\Psi_{hk}).$$

It follows that

$$(9.12)\qquad \begin{aligned} &\Big| \|u\|^2 - \|v_H\|^2 - \sum_{h,k} \|w_{hk}\|^2 \Big| \\ &\leq \sum_{h=1}^{H} \Big| \|v_{h-1}\|^2 - \|v_h\|^2 - \sum_k \|w_{hk}\|^2 \Big| \\ &\leq 2\gamma' \sum_{h,k} B_u(\Psi_{hk}) \leq 2\gamma' \varepsilon^2 \end{aligned}$$

where the last inequality is taken from lemma 9.3. From the definition of the v_h we get the identity

$$(9.13)\qquad \begin{aligned} \langle v_{h-1} | S v_{h-1} \rangle &= \langle v_h | S v_h \rangle + \sum_k \langle w_{hk} | S w_{hk} \rangle + \\ &+ 2\mathrm{Re} \sum_k \langle v_h | S w_{hk} \rangle + 2\mathrm{Re} \sum_{j<k} \langle w_{hj} | S w_{hk} \rangle . \end{aligned}$$

To estimate $\langle v_h | S w_{hk} \rangle$ let $S=T+V$; we have

$$2 | \langle v_h | T w_{hk} \rangle | \leq \int_{\Psi_{hk}} \{ |v_h|^2 + |T w_{hk}|^2 \} dx \leq 2\gamma' B_u(\Psi_{hk}) \leq 2\gamma' \varepsilon$$

by lemma 9.3. Write $v_h = v_h' + v_h''$, where $v_h' = v_h \Phi_{hko}$ and $v''_h = v_h - v_h'$. By (9.6) we have

$$\mathrm{supp}(v_h'') \subset \complement \Omega(h,k,r',r+1)$$

and hence by (9.1.ii) and (9.7)

$$d(\mathrm{supp}(v_h''),\ \mathrm{supp}(w_{hk})) \geq \rho \geq \rho_\varepsilon .$$

Assumption (i) of the theorem gives

$$| \langle v_h'' | V w_{hk} \rangle | \leq \varepsilon \{ \|v_h''\|^2 + \|T v_h''\|^2 + \|w_{hk}\|^2 + \|T w_{hk}\|^2 \}.$$

By construction we have

$$\|v_h''\| \leq \|v_h\| \quad \text{and}$$

$$\|T v_h''\| \leq \|T v_h\| + 2\gamma \|\mathrm{grad}\ v_h\| + \gamma \|v_h\|$$

because of (9.11); it follows

$$|\langle v_h''|Vw_{hk}\rangle| \le \varepsilon\{(1+4\gamma^2)\|v_h\|^2+8\gamma^2\|\mathrm{grad}\ v_h\|^2$$

$$+\ 4\|Tv_h\|^2+\|w_{hk}\|^2+\|Tw_{hk}\|^2\}$$

$$\le \varepsilon\{(4+8\gamma^2)B_{v_h}(\mathbb{R}^m)+B_{w_{hk}}(\mathbb{R}^m)\}$$

$$\le \varepsilon\gamma'(5+8\gamma^2)B_u(\mathbb{R}^m) \le \varepsilon\gamma'(5+8\gamma^2)C.$$

By definition of v_h', by (9.9') and (9.7) we have

$$\mathrm{supp}(v_h') \subset \overline{\Omega(h,k,r',r)} \setminus \Omega(h,k,r',r+\alpha\rho+2)$$

$$\subset \overline{\Psi(h,k,r',r,r+\alpha\rho+2)} = \Psi_{hk}$$

and hence

$$\|v_h'\|^2 \le B_{v_h}(\Psi_{hk}) \le \gamma' B_u(\Psi_{hk}) \le \gamma'\varepsilon^2$$

by lemma 9.3. Since V is T-bounded, the estimate

$$\|Vw_{hk}\|^2 \le a\{\|w_{hk}\|^2+\|Tw_{hk}\|^2\} \le aB_{w_{hk}}(\mathbb{R}^m) \le a\gamma' C$$

holds with a constant a depending on T and V only. It follows

$$|\langle v_h'|Vw_{hk}\rangle| \le \{\gamma'\varepsilon^2 a\gamma' C\}^{1/2} = C_1\varepsilon$$

where C_1 is independent of ε and u. It remains to estimate $\langle w_{hj}|Sw_{hk}\rangle$ for $j \neq k$. By (9.8) the supports of w_{hj} and w_{hk} have distance $\ge \alpha^{-1}(2\alpha\rho+2) > 2\rho > \rho_\varepsilon$ from each other, hence by assumption (i)

$$|\langle w_{hj}|Sw_{hk}\rangle| = |\langle w_{hj}|Vw_{hk}\rangle|$$

$$\le \varepsilon\{\|w_{hj}\|^2+\|Tw_{hj}\|^2+\|w_{hk}\|^2+\|Tw_{hk}\|^2\}$$

$$\le 2\varepsilon\gamma' B_u(\mathbb{R}^m) \le 2\varepsilon\gamma' C.$$

Putting all these estimates into (9.13) we get

$$|\langle v_{h-1}|Sv_{h-1}\rangle - \langle v_h|Sv_h\rangle - \sum_k \langle w_{hk}|Sw_{hk}\rangle| \leq C_2\varepsilon$$

and finally by summation with respect to h

(9.14) $$|\langle u|Su\rangle - \langle v_H|Sv_H\rangle - \sum_{h,k} \langle w_{hk}|Sw_{hk}\rangle| \leq C_3\varepsilon$$

where C_2 and C_3 are independent of ε and u.

c) Apply the decomposition of part b) to every element of (u_n): Let $v_h^{(n)}$ and $w_{hk}^{(n)}$ be the components of u_n. By construction we have $|v_H^{(n)}(x)| \leq |u_n(x)|$ for $x \in \mathbb{R}^m$; by (9.10) the support of $v_H^{(n)}$ is contained in the compact set $K(r')$; hence

$$\|v_H^{(n)}\|^2 \leq \int_{K(r')} |u_n(x)|^2 dx.$$

Now $u_n \rightharpoonup 0$ and $Su_n \rightharpoonup 0$ implies $Tu_n \rightharpoonup 0$, because T is S-bounded and hence $\varphi u_n \to 0$ by lemma 3.1 and lemma 3.10 for every $\varphi \in C_o^\infty(\mathbb{R}^m)$. This implies $v_H^{(n)} \to 0$ for $n \to \infty$. (9.11') gives

$$\|v_H^{(n)}\|^2 + \|Tv_H^{(n)}\|^2 \leq \gamma' B_u(\mathbb{R}^m) \leq \gamma' C;$$

hence the sequence $(Sv_H^{(n)})$ is bounded and $\langle v_H^{(n)}|Sv_H^{(n)}\rangle \to 0$ for $n \to \infty$. The construction of $w_{hk}^{(n)}$ and (9.7) show that $\mathrm{supp}(w_{hk}^{(n)})$ is contained in $\overline{\Omega(h,k,r',r+\alpha\rho+1)}$ where r depends on n and satisfies $r+\alpha\rho+2 < r' \leq 2r$ according to lemma 9.3. It follows

$$\mathrm{supp}(w_{hk}^{(n)}) \subset \overline{\Omega(h,k,r',r'/2+\alpha\rho+1)} \subset \Omega(h,k,r',r'/2)$$

for all n,h and k. We have $r' > \rho+2$ by lemma 9.3 and $\rho \geq r_\varepsilon$ by construction; hence $s=r'/2$ satisfies $r_\varepsilon < r' = 2s$ and $s \leq r'-1$. Assumption (ii) of the theorem gives

$$|\langle w_{hk}^{(n)}|V_{hk}w_{hk}^{(n)}\rangle| \leq \varepsilon\|w_{hk}^{(n)}\| \{\|w_{hk}^{(n)}\| + \|Tw_{hk}^{(n)}\|\}$$

$$\leq 2\varepsilon B_{w_{hk}^{(n)}}(\mathbb{R}^m) \leq 2\varepsilon\gamma' C.$$

By definition of $\eta_{hk}=\min\,\sigma(S_{hk})$ we have

$$\langle u|S_{hk}u\rangle \geq \eta_{hk}\|u\|^2 \quad \text{for all } u \in C_o^\infty(\mathbb{R}^m);$$

hence

$$\begin{aligned}\langle w_{hk}^{(n)}|Sw_{hk}^{(n)}\rangle &= \langle w_{hk}^{(n)}|S_{hk}w_{hk}^{(n)}\rangle + \langle w_{hk}^{(n)}|V_{hk}w_{hk}^{(n)}\rangle \\ &\geq \eta_{hk}\|w_{hk}^{(n)}\|^2 - 2\varepsilon\gamma' C.\end{aligned} \tag{9.15}$$

Applying (9.14) to u_n we get

$$\begin{aligned}\langle u_n|Su_n\rangle - \langle v_H^{(n)}|Sv_H^{(n)}\rangle &\geq \sum_{h,k}\langle w_{hk}^{(n)}|Sw_{hk}^{(n)}\rangle - C_3\varepsilon \\ &\geq \sum_{h,k}\eta_{hk}\|w_{hk}^{(n)}\|^2 - C_4\varepsilon \\ &\geq \eta\sum_{h,k}\|w_{hk}^{(n)}\|^2 - C_4\varepsilon \\ &\geq \eta\{1-\|v_H^{(n)}\|^2\} - C_5\varepsilon\end{aligned}$$

where the last inequality follows from (9.12) applied to u_n. For $n \to \infty$ we have $\langle v_H^{(n)}|Sv_H^{(n)}\rangle \to 0$ and $\|v_H^{(n)}\| \to 0$, hence

$$\lambda = \lim_{n\to\infty}\langle u_n|Su_n\rangle \geq \eta - C_5\varepsilon$$

which proves $\lambda \geq \eta$, q.e.d.

10. The essential spectrum of the Hamiltonian of an N-particle system with external forces

Let S be the Hamiltonian of an N-particle system with external forces, i.e. not all operators $V_{j_1 \dots j_n}$ vanish. By Γ we denote the rotation symmetry group and by Π the permutation symmetry group of S; both have been defined in section 7. If D and E are irreducible unitary representations of Γ and Π respectively, then $D\otimes E$ is an irreducible unitary representation of $\Gamma\times\Pi$. Let P_{DE} be the corresponding projection, and assume that $P_{DE}\neq 0$ so that the reducing subspace $P_{DE}L_2(\mathbb{R}^{3N})$ for S is not trivial. Our aim is to find the essential spectrum of the reduced operator $S(D,E)=S\,|\,P_{DE}C_o^\infty(\mathbb{R}^{3N})$.

Let $Z=(Z_1,Z_2)$ be a decomposition of the system (i.e. of the set $\{1,\dots,N\}$) into two disjoint subsets Z_1,Z_2 such that Z_1 is not empty. Define

$$(10.1)\quad \begin{cases} S_1(Z) = T_{k_1\dots k_m} + \sum\limits_{(j_1\dots j_n)\subset(k_1\dots k_m)} W_{j_1\dots j_n} \\ \qquad \text{for } Z_1 = \{k_1,\dots,k_m\}; \\ S_2(Z) = S_{j_1\dots j_n} \qquad \text{if } Z_2=\{j_1,\dots,j_n\} \text{ and} \\ S_2(Z) = 0 \qquad \text{if } Z_2 \text{ is empty;} \\ S(Z) = S_1(Z) + S_2(Z). \end{cases}$$

Hence S(Z) is the operator obtained by removing from S all terms $V_{j_1\dots j_n}$ such that $\{j_1,\dots,j_n\}\cap Z_1\neq\emptyset$ and all terms $W_{j_1\dots j_n}$ such that $\{j_1,\ldots,j_n\} \cap Z_i \neq \emptyset$ for i=1 <u>and</u> i=2.

If $Z_j \neq \emptyset$ for j=1,2, let $L_2(Z_j)$ denote the L_2-space of functions of the variables x_i with $i \in Z_j$. $S_j(Z)$ is an operator in $L_2(\mathbb{R}^{3N})$ acting on the variables in Z_j only; hence there is a unique operator $S_o(Z_j)$ in $L_2(Z_j)$ which generates $S_j(Z)$ in the sense of definition 5.1 and $S_o(Z_j)$ is essentially selfadjoint by theorem 5.2. We consider the group $\Gamma \times \Pi(Z_j)$ of symmetries of $S_o(Z_j)$, where $\Pi(Z_j)$ is the subgroup of Π generated by the transpositions of identical particles in Z_j. Let D_j and E_j be irreducible representations of Γ and $\Pi(Z_j)$ respectively, let P_{D_j} and P_{E_j} be the corresponding projections in $L_2(Z_j)$ and assume $P_{D_j}P_{E_j} \neq 0$. Then $P_{D_j}P_{E_j}L_2(Z_j)$ is a reducing subspace for $S_o(Z_j)$; let $S_o(Z_j,D_j,E_j)$ denote the reduced operator.

Writing $L_2(\mathbb{R}^{3N}) = L_2(Z_1) \hat{\otimes} L_2(Z_2)$ we have

$$S(Z) = S_o(Z_1) \otimes I_2 + I_1 \otimes S_o(Z_2)$$

where I_j is the identity in $L_2(Z_j)$. Assuming $P_{D_j}P_{E_j} \neq 0$ for j=1,2 the subspace

$$P_{D_1}P_{E_1}L_2(Z_1) \hat{\otimes} P_{D_2}P_{E_2}L_2(Z_2)$$

reduces S(Z) and the reduced operator is

$$S(Z,D_1,D_2,E_1,E_2) = S_o(Z_1,D_1,E_1) \otimes I_2 + I_1 \otimes S_o(Z_2,D_2,E_2) \tag{10.2}$$

where I_j is the identity in $P_{D_j}P_{E_j}L_2(Z_j)$.

Define

$$\eta(Z_j,D_j,E_j) = \min \sigma(S_o(Z_j,D_j,E_j)). \tag{10.3}$$

Then, since $S_o(Z_1)$ is the Hamiltonian of a free system, we have

$\sigma(S_o(Z_1,D_1,E_1))=[\eta(Z_1,D_1,E_1),\infty[$ by theorem 8.10, and hence the spectrum of the operator (10.2) is a half-line

$$[\eta(Z_1,D_1,E_1) + \eta(Z_2,D_2,E_2),\infty[.$$

Recall from section 7 that the relation $(D_1,D_2)<D$ for irreducible representations of Γ means that D is an irreducible component of the representation $\gamma \longmapsto D_1(\gamma)\otimes D_2(\gamma)$. Also for irreducible representations E,E_1,E_2 of $\Pi,\Pi(Z_1),\Pi(Z_2)$ respectively, $(E_1,E_2)\underset{Z}{\models\!\!<} E$ means that $E_1\otimes E_2$ is an irreducible component of the restriction of the representation E to the subgroup $\Pi(Z)=\Pi(Z_1)\times \Pi(Z_2)$.

Define

$$(10.4)\quad \begin{cases} \eta(Z,D,E) = \inf\{\eta(Z_1,D_1,E_1)+\eta(Z_2,D_2,E_2):(D_1,D_2)<D, \\ \qquad (E_1,E_2)\underset{Z}{\models\!\!<}E,\ P_{D_j}P_{E_j}\neq 0 \quad \text{for} \quad j=1,2\} \\ \qquad \text{if} \quad Z_2\neq\emptyset, \quad \text{and} \\ \eta(Z,D,E) = \eta(Z_1,D,E), \quad \text{if } Z_2=\emptyset. \end{cases}$$

Now we can state the main result of this section:

<u>10.5 Theorem.</u> Let D and E be irreducible representations of Γ and Π respectively such that $P_DP_E\neq 0$. Then

$$\sigma_e(S(D,E)) = [\eta(D,E),\infty[$$

where $\eta(D,E)=\min\{\eta(Z,D,E):Z=(Z_1,Z_2),\ Z_1\neq\emptyset\}$.

<u>Remark.</u> For arbitrary irreducible representations D and E of Γ and Π respectively with $P_DP_E\neq 0$ we have $\eta(D,E)\leq 0$, i.e. $[0,\infty[\subset\sigma_e(S(D,E))$. For N=1 there is only one admissible decomposition ($Z_1=\{1\}$, $Z_2=\emptyset$) and Π is trivial; therefore $S_o(Z_1)=-(2\mu_1)^{-1}\Delta$ and theorem 8.8 implies $\eta(Z_1,D)=\eta(D)=0$. By induction with respect to the number N of particles and using theorem 10.5 the above statement follows.

<u>Proof of theorem 10.5.</u> 1. In order to prove $[\eta(D,E),\infty[\subset\sigma_e(S(D,E))$, it suffices to show that $[\eta(Z,D,E),\infty[\subset\sigma(S(D,E))$ for all $Z=(Z_1,Z_2)$

such that $Z_1 \neq \emptyset$, because this implies $[\eta(Z,D,E),\infty[\subset \sigma_e(S(D,E))$.

The case $Z_2 \neq \emptyset$ is treated first: here we have to show that for all D_1, D_2, E_1, E_2 such that $(D_1,D_2) < D$, $(E_1,E_2) \underset{Z}{\vdash\!\!<} E$ and $P_{D_j} P_{E_j} \neq 0$ for $j=1,2$, and for all $\lambda \geq \eta(Z_1,D_1,E_1)+\eta(Z_2,D_2,E_2)$ we have $\lambda \in \sigma(S(D,E))$. To prove this, let $\lambda_2 = \eta(Z_2,D_2,E_2)$; then we have

$$\lambda_1 = \lambda - \lambda_2 \geq \eta(Z_1,D_1,E_1).$$

Let $Z_1 = \{k_1,\ldots,k_m\}$ $(1 \leq m \leq N-1)$; then by theorem 8.10, for every $\varepsilon > 0$ there exists $\rho > 0$ such that for every $r > 0$ there is a $D_1 \otimes E_1$-generating subspace N of $C_o^\infty(Z_1)$ such that for all $u \in N$ we have

$$\|S_o(Z_1)u - \lambda_1 u\| \leq \varepsilon \|u\|$$

and $u(x)=0$ for

$$\Big(\sum_{i,j=1}^{m} |x_{k_i} - x_{k_j}|^2\Big)^{1/2} \geq \rho$$

and for $|\tilde{x}| \leq r$ where

$$\tilde{x} = \Big(\sum_{j=1}^{m} \mu_{k_j}\Big)^{-1} \sum_{j=1}^{m} \mu_{k_j} x_{k_j}$$

is the center of mass of the particles $k_1,\ldots,k_m$. If $R=r-\rho$ is positive and $(x_{k_1},\ldots,x_{k_m})$ is in the support of $u \in N$, it follows that

$$|x_{k_i}| \geq |\tilde{x}| - |\tilde{x} - x_{k_i}| \geq r - \max_j |x_{k_j} - x_{k_i}| \geq r - \rho = R$$

for $i=1,\ldots,m$. Hence, given $\varepsilon > 0$ and $R > 0$, there exists a $D_1 \otimes E_1$-generating subspace N of $C_o^\infty(Z_1)$ such that for all $u \in N$ we have

$$\|S_o(Z_1)u - \lambda_1 u\| \leq \varepsilon \|u\| \quad \text{and}$$

$$u(x)=0 \quad \text{for} \quad \min\{|x_j| : j \in Z_1\} \leq R.$$

Since $\lambda_2 \in \sigma(S_o(Z_2,D_2,E_2))$ there exists a $D_2\otimes E_2$-generating subspace M of $C_o^\infty(Z_2)$ such that

$$\|S_o(Z_2)v - \lambda_2 v\| \leq \varepsilon\|v\|$$

for all $v \in M$ by theorem 8.11. Let R' be the smallest number such that $\max\{|x_j|: j\in Z_2\} \leq R'$ for all x in the support of v for all $v \in M$. Then $N\otimes M$ is a $D_1\otimes E_1\otimes D_2\otimes E_2$-generating subspace of $C_o^\infty(\mathbb{R}^{3N})$ such that for all $w \in N\otimes M$ we have

$$\|S(Z)w - \lambda w\| \leq 2d^{1/2}\varepsilon\|w\|$$

(compare part b of the proof of theorem 8.10), where d is the dimension of $N\otimes M$, and $w(x)=0$ for all x such that

$$\min\{|x_j|: j\in Z_1\} \leq R \quad \text{or} \quad \max\{|x_j|: j\in Z_2\} \geq R'.$$

It follows that $\|S(Z)w\| \leq C_1\|w\|$ for all $w \in N\otimes M$ with $C_1 = |\lambda| + 2d^{1/2}\varepsilon$.

By (6.12) the operator T of kinetic energy is S(Z)-bounded; hence there is a constant $C_2 > 0$ such that $\|Tw\| \leq C_2\|w\|$ for all $w \in N\otimes M$. By definition of S(Z), the operator S-S(Z) is a sum of operators $V_{j_1\dots j_n}$ such that $\{j_1,\dots,j_n\}\cap Z_1 \neq \emptyset$ and of operators $W_{j_1\dots j_n}$ such that $\{j_1,\dots,j_n\}$ has a non-empty intersection with both Z_1 and Z_2. By (6.2b) and (6.7.b) all these operators are small on $N\otimes M$ if R is taken sufficiently large (in particular we need $R > R'$), more precisely we have

$$\|Sw - S(Z)w\| \leq \varepsilon\|w\| \quad \text{for all} \quad w \in N\otimes M$$

if R is large enough. Hence we have

$$\|Sw - \lambda w\| \leq C_3 \epsilon \|w\|$$

for all $w \in N \otimes M$ with $C_3 = 1 + 2d^{1/2}$.

We now construct a non-trivial subspace of $P_D P_E C_o^{\infty}(\mathbb{R}^{3N})$ on which $S - \lambda I$ is small. $N \otimes M$ is a sum of $D_1 \otimes D_2$-generating subspaces; by definition of the relation $(D_1, D_2) < D$, $N \otimes M$ contains a nontrivial D-generating subspace. Hence $L = P_D(N \otimes M)$ is not $\{0\}$ and is contained in $N \otimes M$. The projections P_D and P_E commute; we are going to show that there is a number $C_4 > 0$ such that

$$\|P_E w\| \geq C_4 \|w\| \quad \text{for all} \quad w \in N \otimes M.$$

Hence

$$K = P_E L = P_D P_E (N \otimes M) \neq \{0\}$$

is a non-trivial subspace of $P_D P_E C_o^{\infty}(\mathbb{R}^{3N})$ such that for all

$$u = P_E w \in K \qquad (w \in L)$$

we have

$$\|(S - \lambda I) u\| = \|P_E (S - \lambda I) w\| \leq$$

$$\|(S - \lambda I) w\| \leq C_3 \epsilon \|w\| \leq \frac{C_3}{C_4} \epsilon \|u\|$$

which proves $\lambda \in \sigma(S(D, E))$.

It remains to prove the existence of $C_4 > 0$ such that

$$\|P_E w\| \geq c_4 \|w\| \quad \text{for all} \quad w \in N \otimes M.$$

If $\pi \in \Pi \backslash (\Pi(Z_1) \times \Pi(Z_2))$ then $\pi^{-1} Z_1 \cap Z_2 \neq \emptyset$, and hence for $w \in N \otimes M$ the functions $U(\pi)w$ and w have disjoint support (note that $w(x) = 0$ for

$$\min \{|x_j| : j \in Z_1\} \leq R \quad \text{and for} \quad \max \{|x_j| : j \in Z_2\} \geq R'$$

and that $R > R'$); consequently

$$\langle w | U(\pi) w \rangle = 0.$$

By definition (7.7) of P_E it follows $\langle w | P_E w \rangle = \langle w | Q_E w \rangle$ with

$$Q_E = \frac{d(E)}{|\Pi|} \sum_{\pi \in \Pi(Z_1) \times \Pi(Z_2)} \chi_E(\pi)^* U(\pi)$$

where $d(E)$ and χ_E are the dimension and the character of E respectively and where $|\Pi|$ is the order of Π. The relation $(E_1, E_2) \underset{Z}{\prec} E$ means that $E | \Pi(Z_1) \times \Pi(Z_2)$ is unitarily equivalent to a sum $\bigoplus_{i=0}^{k} n_i E^{(i)}$ with integers $n_i > 0$ and irreducible representations $E^{(i)}$ of $\Pi(Z_1) \times \Pi(Z_2)$ such that $E^{(0)} = E_1 \otimes E_2$. For the corresponding characters this implies

$$\chi_E(\pi) = \sum_{i=0}^{k} n_i \chi^{(i)}(\pi) \quad \text{for all } \pi \in \Pi(Z_1) \times \Pi(Z_2).$$

It follows that

$$Q_E = \sum_{i=0}^{k} \nu_i P_i \quad \text{with} \quad \nu_i > 0$$

and P_i the projection corresponding to $E^{(i)}$. By construction we have

$$P_0 w = w \quad \text{for all} \quad w \in N \otimes M,$$

hence

$$P_i w=0 \quad \text{for} \quad i=1,2,\dots,k$$

and consequently

$$\langle w|P_E w\rangle=\langle w|Q_E w\rangle=\nu_0\|w\|^2$$

which implies

$$\|P_E w\| \geq \nu_0\|w\|.$$

To complete part 1 of the proof, we have to consider the case $Z_2=\emptyset$, $Z_1=\{1,\dots,N\}$. Here we have to show that $\lambda\in\sigma(S(D,E))$ for all $\lambda \geq \eta(Z_1,D,E)$. By theorem 8.10, for every $\varepsilon > 0$ and $R > 0$ there exists a $D\otimes E$-generating subspace N of $C_o^\infty(\mathbb{R}^{3N})$ such that for all $u \in N$ we have

$$\|S_1(Z)u-\lambda u\| \leq \varepsilon\|u\| \quad \text{and}$$

$$u(x)=0 \quad \text{for} \quad \min\{|x_j|: j=1,\dots,N\} \leq R.$$

Now

$$S-S_1(Z)=\sum_{(j_1\dots j_n)\subset(1\dots N)} V_{j_1\dots j_n};$$

by (6.2.b) we have

$$\|Su-S_1(Z)u\| \leq \varepsilon\|u\| \quad \text{for all} \quad u \in N$$

if R is sufficiently large. Hence

$$\|Su-\lambda u\| \leq 2\varepsilon\|u\| \quad \text{for all} \quad u \in N,$$

which implies $\lambda\in\sigma(S(D,E))$.

2. The second part of the proof has to show that $\sigma_e(S(D,E)) \subset [\eta(D,E),\infty[$; this is done with the help of theorem 9.4. In order to apply this theorem, we have to define the functions $p_{hk\ell r}$, in (9.0).

Let $Z^{hk}=(Z_1^{hk},Z_2^{hk})$ be a numbering of all decompositions $Z=(Z_1,Z_2)$ of $\{1,\dots,N\}$ such that $Z_1\neq\emptyset$, where $h\in\{1,\dots,N\}$ is 1 plus the number of elements of Z_2^{hk}, and where $k\in\{1,\dots,k(h)\}$ numbers the different decompositions for fixed h. For every pair (h,k) and every $r'>0$ let $p_{hk\ell r'}$, $\ell\in\{1,\dots,\ell(h,k)\}$ be a numbering of the functions with values

$$(10.6)\qquad \begin{cases} p(x_i)-(h-1)r' & i\in Z_1^{hk} \\ p(x_i)-p(x_j) & i\in Z_1^{hk},\ j\in Z_2^{hk} \\ hr'-p(x_j) & j\in Z_2^{hk} \end{cases}$$

where $p(z)=(1+|z|^2)^{1/2}$ for $z\in\mathbb{R}^3$. Define the sets $\Omega(h,k,r',r)$ by (9.0); then (9.1) and (9.2) hold. The proof of this fact is contained in lemma 10.8 below.

It follows from (6.2.c) and (6.7.c) that assumption (i) of theorem (9.4) is satisfied. For every pair (h,k) we define the operator

$$V_{hk}=S-S(Z^{hk})$$

where S(Z) is given by (10.1); then assumption (ii) of theorem (9.4) is satisfied: Let $\varepsilon>0$ be given and choose r_ε according to (6.2.b) and (6.7.b) simultaneously. V_{hk} is the sum of all operators

$$V_{j_1\dots j_n}\quad\text{such that}\quad \{j_1,\dots,j_n\}\cap Z_1^{hk}\neq\emptyset$$

and of all operators

$$W_{j_1\dots j_n}\quad\text{such that } \{j_1,\dots,j_n\}\cap Z_i^{hk}\neq\emptyset\quad\text{both for } i=1 \text{ and } i=2$$

We use the following elementary inequalities valid for $p(z)=(1+|z|^2)^{1/2}$:

$$(10.7)\qquad \begin{array}{ll} \text{(i)} & p(x\pm y)\le p(x)+|y|\le p(x)+p(y) \\ \text{(ii)} & |x-y|\ge p(x)-p(y) \\ \text{(iii)} & p(x)\ge r \Rightarrow |x|\ge r-1. \end{array}$$

Let $2r_\varepsilon+2< r'\leq 2s\leq 2r'-2$ and $x\in \Omega(h,k,r',s)$. For $i\in Z_1^{hk}$ we have $p(x_i)>(h-1)r'+s \geq s > r_\varepsilon+1$ by (9.0) and (10.6), hence $|x_i|> r_\varepsilon$ by (10.7.iii); it follows from (6.2.b) that

$$\|V_{j_1\ldots j_n}u\| \leq \varepsilon(\|u\|+\|Tu\|)$$

for $\{j_1,\ldots,j_n\} \cap Z_1^{hk}\neq\emptyset$ and for all $u\in C_o^\infty(\mathbb{R}^{3N})$ such that $\mathrm{supp}(u) \subset \Omega(h,k,r',s)$. For $i\in Z_1^{hk}$, $j\in Z_2^{hk}$ we have $p(x_i)-p(x_j)> s > r_\varepsilon$ by (9.0) and (10.6), hence $|x_i-x_j|> r_\varepsilon$ by (10.7.iii); it follows from (6.7.b) that

$$\|W_{j_1\ldots j_n}u\| \leq \varepsilon(\|u\|+\|Tu\|)$$

for all $u \in C_o^\infty(\mathbb{R}^{3N})$ such that $\mathrm{supp}(u)\subset \Omega(h,k,r',s)$ if $\{j_1\ldots j_n\} \cap Z_1^{hk}\neq\emptyset$ and $\{j_1\ldots j_n\} \cap Z_2^{hk}\neq\emptyset$. Since there are less then 2^{N+1} terms in V_{hk}, we have

$$\|V_{hk}u\| \leq 2^{N+1}\varepsilon(\|u\|+\|Tu\|)$$

for all $u \in C_o^\infty(\mathbb{R}^m)$ such that $\mathrm{supp}(u)\subset \Omega(h,k,r',s)$; redefining ε and r_ε in a suitable way we get (ii) of theorem 9.4.

If both groups Γ and Π consist of their neutral element only, then there exist no representations except D=1, E=1; hence $P_DP_E=I$, and theorem 9.4 implies the result because $S_{hk}=S(Z^{hk})$ for all h,k.

In the general case, however, we have to modify theorem 9.4 in the following way:

We replace $C_o^\infty(\mathbb{R}^{3N})$ by $P_DP_EC_o^\infty(\mathbb{R}^{3N})=D(S(D,E))$.

By definition (10.6) all functions $p_{hk\ell r'}$ are invariant under Γ; hence the same is true for the functions Φ_{hks} defined by (9.5). The permutations $\pi \in \Pi$ obviously permute the functions $p_{hk\ell r'}$, more precisely $p_{hk\ell r'}(x)=p_{hk'\ell' r'}(\pi x)$, where $Z^{hk'}=\pi Z^{hk}$, hence $\Phi_{hks}(x)=\Phi_{hk's}(\pi x)$, and consequently

$$\sum_{k=1}^{k(h)} \Phi_{hks} \quad \text{is invariant under } \Pi.$$

It follows that for $u \in D(S(D,E))$ the functions v_h defined by (9.9) are also in $D(S(D,E))$. By lemma 7.5 and by (7.13) we get

$$D(S(D,E)) \subset \bigoplus_i D(S(Z^{hk}, D_1^{(i)}, D_2^{(i)}, E_1^{(i)}, E_2^{(i)})) \qquad \text{for } h > 1,$$

where the sum is taken over all $(D_1^{(i)}, D_2^{(i)}) < D$ and all $(E_1^{(i)}, E_2^{(i)}) \underset{Z^{hk}}{\models\!\!<} E$. For $\pi \in \Pi(Z_1^{hk}) \times \Pi(Z_2^{hk})$ we have $\Phi_{hks}(x) = \Phi_{hks}(\pi x)$; hence the functions w_{hk} defined by (9.9) are in

$$\bigoplus_i D(S(Z^{hk}, D_1^{(i)}, D_1^{(i)}, E_1^{(i)}, E_2^{(i)}))$$

and consequently the inequality (9.15) holds with η_{hk} replaced by the number $\eta(Z^{hk}, D, E)$ defined in theorem 10.5 (note that $S_{hk} = S(Z^{hk})$). For $h=1$ this result is trivial. This completes the proof of theorem 10.5.

The following lemma has been used in the proof:

10.8 Lemma. Let $\Omega(h,k,r',r)$ be defined by (9.0) with the functions $p_{hk\ell r'}$ defined by (10.6). Then the assertion (9.1) holds with $\alpha = 2$ and (9.2) holds with

$$K(r') = \{x : x \in R^{3N},\ |x_i| \leq Nr' \quad \text{for } i=1,\dots,N\}.$$

Proof. 1. (Proof of (9.1.i)). For $h=1$ there is only one decomposition Z^{11}, hence we may take $h \geq 2$. Let $k \neq k'$, $s > 0$, $t > 0$, $r' > 0$ and $s+t-r' = \rho > 0$; then there exists $i \in Z_1^{hk} \cap Z_2^{hk'}$. For $x \in \Omega(h,k,r',s)$ and $y \in \Omega(h,k',r',t)$ it follows that

$$p(x_i) - (h-1)r' > s \quad \text{and} \quad hr' - p(y_i) > t.$$

By (10.7.ii) we get

$$|x-y| \geq |x_i - y_i| \geq p(x_i) - p(y_i)$$
$$> (h-1)r' + s - hr' + t \geq \rho > \tfrac{1}{2}\rho, \qquad \text{q.e.d.}$$

2. (Proof of (9.1.ii)). Let $s > 0$, $t > 0$, $r' > 0$, $s-t=\rho > 0$, $x\in \Omega(h,k,r',s)$ and $y\in \Omega(h,k,r',t)$. If $h=1$, then there exists $i\in\{1,\dots,N\}$ such that

$$p(y_i) \leq t, \quad \text{whereas} \quad p(x_i) > s;$$

hence

$$|x-y| \geq |x_i-y_i| \geq p(x_i)-p(y_i) > s-t=\rho > \tfrac{1}{2}\rho.$$

If $h \geq 2$, then there exist $i\in Z_1^{hk}$ and $j\in Z_2^{hk}$ such that at least one of the following inequalities holds:

$$\text{a)} \qquad p(y_i)-(h-1)r' \leq t,$$

$$\text{b)} \qquad p(y_i)-p(y_j) \leq t,$$

$$\text{c)} \qquad hr'-p(y_j) \leq t.$$

In case a) we also have $p(x_i)-(h-1)r'> s$ and hence

$$|x-y| \geq |x_i-y_i| \geq p(x_i)-p(y_i) > s-t=\rho > \tfrac{1}{2}\rho.$$

In case b) we also have $p(x_i)-p(x_j)> s$ and hence

$$|x-y| \geq \max \{|x_i-y_i|, |y_j-x_j|\}$$

$$\geq \tfrac{1}{2}(p(x_i)-p(y_i)+p(y_j)-p(x_j)) > \tfrac{1}{2}(s-t)=\tfrac{1}{2}\rho.$$

In case c) we also have $hr'-p(x_j)> s$ and hence

$$|x-y| \geq |y_j-x_j| \geq p(y_j)-p(x_j) > s-t=\rho > \tfrac{1}{2}\rho.$$

In any case we get $|x-y| > \tfrac{1}{2}\rho$, q.e.d.

3. (Proof of (9.2)). Let $0< r< r'$ and $x\in \complement(\bigcup_{h,k} \Omega(h,k,r',r))$, hence $x\in \complement\Omega(h,k,r',r)$ for all h,k. For $h=1$ there exists $i\in\{1,\dots,N\}$ such that $p(x_i) \leq r$; for $h \geq 2$ there exist $i\in Z_1^{hk}$ and $j\in Z_2^{hk}$ such that at least one of the following inequalities holds:

a) $p(x_i)-(h-1)r' \leq r$,

b) $p(x_i)-p(x_j) \leq r$,

c) $hr'-p(x_j) \leq r$.

Starting with h=1 we find $j_1 \in \{1,\dots,N\}$ such that $p(x_{j_1}) \leq r$. We apply induction with respect to h: Let $2 \leq h \leq N$ and suppose there are h-1 different integers $j_i \in \{1,\dots,N\}$ such that

$$p(x_{j_i}) \leq (h-2)r'+r \quad \text{for } i=1,\dots,h-1.$$

Choose Z^{hk} such that $Z_2^{hk}=\{j_1,\dots,j_{h-1}\}$; then

$$hr'-p(x_{j_i}) \geq 2r'-r > r,$$

so that c) is not true for Z^{hk}. Hence there exists $i \in Z_1^{hk}$ such that either

$$p(x_i)-(h-1)r' \leq r \quad \text{or} \quad p(x_i)-p(x_j) \leq r \quad \text{for some } j \in Z_2^{hk};$$

in the first case

$$p(x_i) \leq (h-1)r'+r,$$

in the second case

$$p(x_i) \leq p(x_j)+r \leq (h-2)r'+2r < (h-1)r'+r.$$

If we take $i=j_h$, the induction is complete. For h=N we get

$$|x_j| < p(x_j) < Nr' \quad \text{for } j=1,\dots,N, \quad \text{q.e.d.}$$

For atomic systems with fixed nucleus we have the following simpler result.

10.9 Theorem. If in theorem 10.5 all the interaction operators $W_{j_1 \dots j_n}$ are non-negative, we have

$$\sigma_e(S(D,E)) = [\eta_1(D,E), \infty[,$$

where

$$\eta_1(D,E) = \min \{\eta(Z,D,E) : Z=(Z_1,Z_2),\ |Z_1| = 1\}.$$

Proof. In this case for any decomposition Z the operator $S_o(Z_1)$ is non-negative and the remark following theorem 10.5 implies $\eta(Z_1,D_1,E_1) = 0$. Therefore

$$\eta(D,E) = \inf \Big\{\eta(Z_2,D_2,E_2) : Z=(Z_1,Z_2),\ Z_1 \neq \emptyset,$$

$$\text{there exist } D_1 \text{ and } E_1 \text{ such that}$$

$$(10.10) \qquad (D_1,D_2) < D,\ (E_1,E_2) \underset{Z}{\models\!\!<} E,\ P_{E_j} P_{D_j} \neq 0$$

$$\text{for } j=1,2\Big\}$$

and it suffices to show that for every decomposition $Z=(Z_1,Z_2)$ with $Z_1 \neq \emptyset$ and representations D_i, E_i with $(D_1,D_2)<D$, $(E_1,E_2) \underset{Z}{\models\!\!<} E$ and $P_{E_j} P_{D_j} \neq 0$ there exist a decomposition $\hat{Z}=(\hat{Z}_1,\hat{Z}_2)$ with $|\hat{Z}_1|=1$ and representations $\hat{D}_i$ of Γ and $\hat{E}$ of $\Pi(\hat{Z}_2)$ such that

$(\hat{D}_1,\hat{D}_2)<D$,

$\hat{E} \underset{\hat{Z}}{\models\!\!<} E$, i.e. $\hat{E}$ is contained in the restriction of E to $\Pi(\hat{Z}_2)$,

$P_{\hat{D}_1} \neq 0$ in $L_2(\hat{Z}_1)$ and $P_{\hat{E}} P_{\hat{D}_2} \neq 0$ in $L_2(\hat{Z}_2)$,

$\eta(\hat{Z}_2,\hat{D}_2,\hat{E}) \leq \eta(Z_2,D_2,E_2)$.

Let $\hat{Z}$ be any decomposition with $\hat{Z}_1 \subset Z_1$ and $|\hat{Z}_1|=1$. By lemma 7.5 there exist representations $\hat{D}_1$ and D_3 of Γ such that $(\hat{D}_1,D_3)<D_1$, $P_{\hat{D}_1} \neq 0$ in $L_2(\hat{Z}_1)$ and $P_{D_3} \neq 0$ in $L_2(Z_1 \setminus \hat{Z}_1)$. Then $(\hat{D}_1,D_3,D_2)<D$ and by lemmas 7.5

and 7.6 there exists a representation $\hat{D}_2$ of Γ such that $P_{\hat{D}_2}\neq 0$ in $L_2(\hat{Z}_2)$, $(D_3,D_2)<\hat{D}_2$ and $(\hat{D}_1,\hat{D}_2)<D$. Furthermore by lemma 7.12 there exist representations E_3 of $\Pi(Z_1\cap\hat{Z}_2)$ and $\hat{E}$ of $\Pi(\hat{Z}_2)$ such that

$$\hat{E} \underset{\hat{Z}}{\models} E,\ (E_3,E_2) \underset{(Z_1\cap\hat{Z}_2,Z_2)}{\models} \hat{E}.$$

Hence, by lemma (10.11) below (with N replaced by N-1), we have $P_{\hat{D}_2}P_{\hat{E}}\neq 0$ and from theorem 10.5 and (10.10) follows

$$\eta(\hat{Z}_2,\hat{D}_2,\hat{E}) = \min\sigma(S_o(\hat{Z}_2,\hat{D}_2,\hat{E}))$$
$$\leq \min\sigma_e(S_o(\hat{Z}_2,\hat{D}_2,\hat{E})) \leq \eta(Z_2,D_2,E_2).$$

<u>10.11 Lemma.</u> Let D and E be irreucible unitary representations of Γ and Π respectively with $P_D\neq 0$ in $L_2(\mathbb{R}^{3N})$. Then

$$P_EP_D \neq 0.$$

<u>Proof</u>. Since $P_D\neq 0$, by lemma 7.5 it follows that there exist representations $D_1,\dots,D_N$ with

$$P_{D_i} \neq 0 \quad \text{in } L_2(\mathbb{R}^3), \quad (D_1,\dots,D_N) < D.$$

Now choose mutually orthogonal D_i-generating subspaces M_i of $L_2(\mathbb{R}^3)$ and let

$$M := M_1 \otimes M_2 \otimes\dots\otimes M_N \subset L_2(\mathbb{R}^{3N}).$$

This is possible because $M_i\perp M_j$ if $D_i\neq D_j$ and because for every i there are infinitely many mutually orthogonal D_i-generating subspaces of $L_2(\mathbb{R}^3)$ (comoare the proof of theorem 8.8), so that in case $D_i=D_j$ we can again choose $M_i\perp M_j$. By definition of the relation $(D_1,\dots,D_N)<D$ it follows that M contains a D-generating subspace M_o. For every $f\in M_o$, and for every pair $\pi_1,\pi_2\in\Pi$ such that $\pi_1\neq\pi_2$, we have $U(\pi_1)f\perp U(\pi_2)f$

and therefore

$$P_E f = \frac{d_E}{|\Pi|} \sum_{\pi \in \Pi} \chi_E(\pi)^* U(\pi) f \neq 0 \quad \text{for } f \in M_o,\ f \neq 0.$$

This implies $P_E P_D \neq 0$.

11. The essential spectrum of the internal Hamiltonian of a free system

Let S be the Hamiltonian of a free N-particle system, and let $\hat{S}$ be its internal Hamiltonian defined by (6.10). By theorem 8.10 the spectrum of S in any non-trivial subspace $P_D P_E L_2(\mathbb{R}^{3N})$ is a half-line $[\eta_{DE}, \infty[$. In this section we consider $\hat{S}$; by theorem 6.11, $\hat{S}$ can be reduced to an operator S_o in $L_2(\mathbb{R}^{3N-3})$ by introducing internal coordinates. The spectrum of S_o is not a half-line in general: this is most easily seen by considering the case N=2, where S_o is of the form

$$-\frac{1}{2\nu}\Delta + W_{12} \text{ in } L_2(\mathbb{R}^3),$$

and where W_{12} is Δ-small at infinity. However, by theorem 10.5 we have

$$\sigma_e(S_o, D) = [0, \infty[$$

for every irreducible representation D of the symmetry-group Γ of S_o such that $P_D \neq 0$. In this case the symmetry group Π consists of at most two elements, the identity and possibly the transposition of the two particles. But this transposition is the same as the reflection of the internal coordinates and can be considered as an element of Γ; hence we need not consider the group Π. Therefore we shall assume $N > 2$ in the sequel and we shall show that the essential spectrum of the restriction of S_o to any of its symmetry subspaces is a half-line.

Let $Z=(Z_1,\ldots,Z_m)$ be a decomposition of the set $\{1,\ldots,N\}$ into m mutually disjoint non-empty subsets, $2 \leq m \leq N$. Denote by $|Z_i|$ the number of elements in Z_i; for convenience assume that $|Z_1| \geq |Z_2| \geq \ldots \geq |Z_m|$

and identify with Z all decompositions Z' obtained from Z by renumbering the subsets Z_i. Let $\hat{S}(Z)$ be the operator obtained by removing from $\hat{S}$ all terms $W_{j_1 \dots j_m}$ corresponding to interactions between particles belonging to different subsets Z_i. If $x \mapsto ax=y$ is a linear coordinate transformation such that $y_1,\dots,y_{N-1}$ are internal coordinates for the entire system, and if the corresponding unitary operator U is defined by (6.4), then by theorem 6.11 the operators $U\hat{S}U^*$ and $U\hat{S}(Z)U^*$ act on the internal coordinates only and are generated in the sense of definition 5.1 by operators S_o and $S_o(Z)$ in $L_2(\mathbb{R}^{3N-3})$. We may ignore the dependence of S_o on the choice of internal coordinates because the operators corresponding to different coordinate systems are unitarily equivalent.

For the case m=2, i.e. for decompositions $Z=(Z_1,Z_2)$, we need an explicit representation of $S_o(Z)$. Let $n=|Z_1| \geq |Z_2|=N-n$ and choose coordinates $z_1,\dots,z_N$ such that $z_2,\dots,z_n$ and $z_{n+1},\dots,z_{N-1}$ are internal coordinates and z_1 and z_N the center-of-mass coordinates for the subsystems Z_1 and Z_2 respectively (if $n=N-1$ the subsystem Z_2 has no internal coordinates; but $N > 2$ implies $n > 1$, so that Z_1 does have internal coordinates). Let ν_1 and ν_2 be the total mass of Z_1 and Z_2 respectively; then $\mu=\nu_1+\nu_2$ is the total mass of the entire system. Now put

$$(11.0) \qquad \begin{cases} y_1 = z_1 - z_N \\ y_j = z_j \quad \text{for } j=2,\dots,N-1 \\ y_N = \mu^{-1}(\nu_1 z_1+\nu_2 z_N). \end{cases}$$

Then $y_1,\dots,y_{N-1}$ are internal coordinates for the entire system. In these coordinates we have

$$(11.1) \qquad L_2(\mathbb{R}^{3N-3}) = L_2(Z_o) \hat{\otimes} L_2(Z_1) \hat{\otimes} L_2(Z_2)$$

if $n < N-1$, where $L_2(Z_o)=L_2(\mathbb{R}^3)$, $L_2(Z_1)=L_2(\mathbb{R}^{3n-3})$, $L_2(Z_2)=L_2(\mathbb{R}^{3N-3n-3})$

correspond to the coordinate y_1 and to the internal coordinates of Z_1 and Z_2 respectively, and

(11.2) $$S_o(Z)=T_o\otimes I_1\otimes I_2+I_o\otimes S_o(Z_1)\otimes I_2+I_o\otimes I_1\otimes S_o(Z_2)$$

where $T_o=-\frac{1}{2\nu}\Delta$ in $L_2(Z_o)$ with $\frac{1}{\nu}=\frac{1}{\nu_1}+\frac{1}{\nu_2}$, $S_o(Z_i)$ in $L_2(Z_i)$ are generating operators for the internal operators $\hat{S}(Z_i)$ of the subsystems, and I_j is the identity in $L_2(Z_j)$ for j=0,1,2. In the case n=N-1 we have

(11.1') $$L_2(\mathbb{R}^{3N-3}) = L_2(Z_o)\ \hat{\otimes}\ L_2(Z_1)$$

and

(11.2') $$S_o(Z) = T_o \otimes I_1+I_o \otimes S_o(Z_1).$$

For the rotation symmetry group Γ of $\hat{S}$ the representation $\gamma\mapsto U(\gamma)$ by unitary operators in $L_2(\mathbb{R}^{3N-3})$ is defined as usual by

$$(U(\gamma)f)(y) = f(\gamma^{-1}y_1,\dots,\gamma^{-1}y_{N-1})$$

in terms of internal coordinates. Hence for the special case (11.1) we have

$$U(\gamma) = U_o(\gamma) \otimes U_1(\gamma) \otimes U_2(\gamma)$$

where the unitary representation $\gamma\mapsto U_i(\gamma)$ in $L_2(Z_i)$ is defined analogically for i=0,1,2. For irreducible representations D_i of Γ we define the corresponding projections P_{D_i} in $L_2(Z_i)$; if $P_{D_i}\neq 0$, the subspace $P_{D_i}L_2(Z_i)$ reduces T_o for i=0 and $S_o(Z_i)$ for i=1,2.

For the permutation symmetry group Π of $\hat{S}$ the representation

$$\pi\mapsto U(\pi) \text{ in } L_2(\mathbb{R}^{3N})$$

was defined in section 7 by

$$(U(\pi)f)(x) = f(\pi^{-1}x)$$

in terms of the original variables, where

$$\pi^{-1}x = (x_{\pi^{-1}(1)},\dots,x_{\pi^{-1}(N)}).$$

Hence in $L_2(\mathbb{R}^{3N-3})$ and in terms of internal coordinates $y=ax$ we have to define $U(\pi)$ by

$$(U(\pi)f)(ax) = f(a\pi^{-1}x) \tag{11.3}$$

for $x \in \mathbb{R}^{3N}$ and $f \in L_2(\mathbb{R}^{3N-3})$. For every decomposition $Z=(Z_1,\dots,Z_m)$ we consider the subgroup $\Pi(Z)$ consisting of all $\pi \in \Pi$ such that $\pi Z_i = Z_i$ for $i=1,\dots,m$. Clearly

$$\Pi(Z) = \Pi(Z_1)\times\dots\times\Pi(Z_m)$$

where $\Pi(Z_i)$ is the subgroup generated by transpositions of identical particles in Z_i.

We need a decomposition of $U(\pi)$ corresponding to a decomposition $Z=(Z_1,Z_2)$. For simplicity let $Z_1=\{1,\dots,n\}$, $Z_2=\{n+1,\dots,N\}$; then the coordinates defined by (11.0) are given by $y=ax$ with a matrix a of the form

$$a = \begin{pmatrix} \frac{\mu_1}{\nu_1} \dots \frac{\mu_n}{\nu_1} & -\frac{\mu_{n+1}}{\nu_2} \dots -\frac{\mu_N}{\nu_2} \\ a_1 & 0 \\ 0 & a_2 \\ \frac{\mu_1}{\mu} \dots \frac{\mu_n}{\mu} & \frac{\mu_{n+1}}{\mu} \dots \frac{\mu_N}{\mu} \end{pmatrix} \tag{11.4}$$

where a_1 and a_2 are $(n-1)\times n$ and $(N-n-1)\times(N-n)$ matrices respectively. It follows from (11.1) and (11.3) that

$$U(\pi_1\times\pi_2) = I_o \otimes U_1(\pi_1) \otimes U_2(\pi_2) \tag{11.5}$$

for $\pi_i \in \Pi(Z_i)$, where I_o is the identity in $L_2(Z_o)$ and where $U_i(\pi_i)$

in $L_2(Z_i)$ is defined by

(11.6) $$(U_i(\pi_i)f)(a_i z) = f(a_i \pi_i^{-1} z)$$

for $z \in \mathbb{R}^{3n}$ or $z \in \mathbb{R}^{3N-3n}$ and for $f \in L_2(Z_i)$. In the case n=N-1 we have $\Pi(Z_2)=\{1\}$ and there is a similar decomposition

(11.5') $$U(\pi_1) = I_o \otimes U_1(\pi_1)$$

corresponding to (11.1').

By definition of Π we have

$$U(\pi)S_o = S_o U(\pi) \quad \text{for all } \pi \in \Pi;$$

similarly, by definition of $S_o(Z)$ and $\Pi(Z)$, we have

$$U(\pi)S_o(Z) = S_o(Z)U(\pi) \quad \text{for all } \pi \in \Pi(Z)$$

and hence

$$U_i(\pi_i)S_o(Z_i) = S_o(Z_i)U_i(\pi_i) \quad \text{for all } \pi_i \in \Pi(Z_i)$$

by (11.2) and (11.5) or by (11.2') and (11.5'). For irreducible representations E of Π and E_i of $\Pi(Z_i)$ define the corresponding projections P_E in $L_2(\mathbb{R}^{3N-3})$ and P_{E_i} in $L_2(Z_i)$; if these operators are not zero, the subspaces $P_E L_2(\mathbb{R}^{3N-3})$ and $P_{E_i} L_2(Z_i)$ reduce S_o and $S_o(Z_i)$ respectively.

Let us now consider the full symmetry group $\Gamma \times \Pi$. For irreducible representations D and E of Γ and Π respectively such that $P_D P_E \neq 0$ define the reduced operator $S_o(D,E)$ in $P_D P_E L_2(\mathbb{R}^{3N-3})$. Similarly for irreducible representations D_i of Γ and E_i of $\Pi(Z_i)$ such that $P_{D_i} P_{E_i} \neq 0$ define the reduced operator $S_o(Z_i, D_i, E_i)$ in $P_{D_i} P_{E_i} L_2(Z_i)$. If D_o is another representation of Γ such that the corresponding projection P_{D_o} in $L_2(Z_o)$ is not zero, the subspace

(11.6) $$P_{D_o}L_2(Z_o) \,\hat{\otimes}\, P_{D_1}P_{E_1}L_2(Z_1) \,\hat{\otimes}\, P_{D_2}P_{E_2}L_2(Z_2)$$

reduces $S_o(Z)$ and the reduced operator is

(11.7) $$\begin{aligned} S_o(Z,D_o,D_1,D_2,E_1,E_2) &= T_o(D_o) \otimes I_1 \otimes I_2 + \\ &+ I_o \otimes S_o(Z_1,D_1,E_1) \otimes I_2 + I_o \otimes I_1 \otimes S_o(Z_2,D_2,E_2), \end{aligned}$$

where $T_o(D_o)$ is the reduced operator for T_o in $P_{D_o}L_2(Z_o)$; similarly, if $|Z_2|=1$, the reducing subspace is

(11.6') $$P_{D_o}L_2(Z_o) \,\hat{\otimes}\, P_{D_1}P_{E_1}L_2(Z_1)$$

and the reduced operator is

(11.7') $$S_o(Z,D_o,D_1,E_1)=T_o(D_o) \otimes I_1+I_o \otimes S_o(Z_1,D_1,E_1).$$

Define

(11.8) $$\eta(Z_i,D_i,E_i) = \min \sigma(S_o(Z_i,D_i,E_i)).$$

Then, since $\sigma(T_o(D_o))=[0,\infty[$ by theorem 8.8, the spectrum of the operators (11.7) and (11.7') is the half-line

$$[\eta(Z_1,D_1,E_1)+\eta(Z_2,D_2,E_2),\infty[\quad \text{and} \quad [\eta(Z_1,D_1,E_1),\infty[$$

respectively.

It remains to discuss the special case where $n=2N$ is even, $|Z_1|=|Z_2|=n$ and where the subsystems Z_1 and Z_2 are <u>identical</u> in the sense that there exists a permutation $\tau \in \Pi$ such that $\tau Z_1=Z_2$ and $\tau Z_2=Z_1$. For simplicity assume that $Z_1=\{1,\dots,n\}$ and

(11.9) $$\tau = \begin{pmatrix} 1,\dots,n, & n+1,\dots,N \\ n+1,\dots,N, & 1,\dots,n \end{pmatrix}.$$

Then $S_o(Z)$ is invariant under τ, and we have to consider the group

$$\hat{\Pi}(Z) = \Pi(Z) \cup \tau\Pi(Z)$$

instead of $\Pi(Z)$. We use the coordinates (11.4) where we now assume $a_1=a_2$, i.e. we choose internal coordinates in the same way for both subsystems. Then the mapping

$$J: L_2(Z_1) \to L_2(Z_2)$$

defined by

(11.9')
$$(Jf)(y_{n+1},\dots,y_{N-1}) = f(y_{n+1},\dots,y_{N-1})$$

for $f \in L_2(Z_1)$ is an isometric isomorphism of $L_2(Z_1)$ onto $L_2(Z_2)$ such that

(11.9")
$$S_o(Z_2) = JS_o(Z_1)J^{-1}.$$

For $\pi \in \Pi(Z)$ the unitary operator $U(\pi)$ is given by (11.5) as before. It follows from (11.3), (11.4) and (11.9) that $U(\tau)$ is given by

(11.10)
$$U(\tau)f = U_o(\tau)f_o \otimes J^{-1}f_2 \otimes Jf_1$$

for $f=f_o\otimes f_1\otimes f_2$, where $U_o(\tau)$ in $L_2(Z_o)$ is defined by

(11.10')
$$(U_o(\tau)f_o)(z) = f_o(-z)$$

for $z \in \mathbb{R}^3$ and $f_o \in L_2(Z_o)$.

Let $E_1^{(i)}$, $i=1,\dots,p$ be all irreducible representations of $\Pi(Z_1)$; denote by d_i and $\chi^{(i)}$ the dimension and the character of $E_1^{(i)}$ respectively. The mapping $\pi \mapsto \tau\pi\tau=\pi'$ is an isomorphism of $\Pi(Z_2)$ onto $\Pi(Z_1)$; hence

$$E_2^{(i)}(\pi) = E_1^{(i)}(\pi') \quad \text{for all } \pi \in \Pi(Z_2)$$

defines an irreducible representation $E_2^{(i)}$ of $\Pi(Z_2)$ of dimension d_i and character $\chi^{(i)}(\pi')$ for $i=1,\dots,p$; and this are all irreducible representations of $\Pi(Z_2)$. It follows that $E_1^{(i)} \otimes E_2^{(j)}$, $i,j=1,\dots,p$ are all irreducible representations of $\Pi(Z)$.

<u>11.11 Lemma.</u> The group $\hat{\Pi}(Z)$ has exactly $\frac{p(p-1)}{2}+2p$ irreducible representations $E^{(ij)}$, $i<j$, and $E_{\pm}^{(i)}$, $i,j \in \{1,\dots,p\}$ such that the corresponding projections in $L_2(Z_o) \hat{\otimes} L_2(Z_1) \hat{\otimes} L_2(Z_2)$ are given by

$$P^{(ij)}=I_o \otimes P_1^{(i)} \otimes P_2^{(j)}+I_o \otimes P_1^{(j)} \otimes P_2^{(i)},$$

where $P_k^{(i)}$ is the projection corresponding to $E_k^{(i)}$ in $L_2(Z_k)$, and

$$P_{\pm}^{(i)}= \frac{1}{2}I_o \otimes P_1^{(i)} \otimes P_2^{(i)} \pm \frac{1}{2}U_o(\tau) \otimes Q^{(i)}(P_1^{(i)} \otimes P_2^{(i)})$$

where the operator $Q^{(i)}$ in $P_1^{(i)}L_2(Z_1)\hat{\otimes}P_2^{(i)}L_2(Z_2)$ is defined as follows: Let u_{jk}, $j=1,\dots,d_i$, $k=1,2,\dots$ be an orthonormal basis of $L_2(Z_1)$ such that

$$U_1(\pi)u_{jk}= \sum_{r=1}^{d_i} E_1^{(i)}(\pi)_{rj}u_{rk}$$

for all j,k and $\pi \in \Pi(Z_1)$. Then

$$Q^{(i)}\left(u_{jk} \otimes Ju_{rs}\right)= u_{js} \otimes Ju_{rk} \quad \text{for all } j,k,r,s,$$

where J is the mapping defined by (11.9').

<u>Proof</u>. Define $E^{(ij)}$ by

$$E^{(ij)}(\pi_1 \times \pi_2) = \begin{pmatrix} E_1^{(i)}(\pi_1) \otimes E_2^{(j)}(\pi_2) & 0 \\ 0 & E_1^{(j)}(\pi_1) \otimes E_2^{(i)}(\pi_2) \end{pmatrix}$$

$$E^{(ij)}(\tau(\pi_1\times\pi_2)) = \begin{pmatrix} 0 & E_1^{(j)}(\pi_1)\otimes E_2^{(i)}(\pi_2) \\ E_1^{(i)}(\pi_1)\otimes E_2^{(j)}(\pi_2) & 0 \end{pmatrix};$$

for $i < j$, $E^{(ij)}$ is an irreducible representation of $\hat{\Pi}(Z)$ of dimension $2d_i d_j$ and equivalent to $E^{(ji)}$, whereas the different $E^{(ij)}$ for $i < j$ are inequivalent (even over $\Pi(Z)$). Next define $E_{\pm}^{(i)}$ by

$$E_{\pm}^{(i)}(\pi_1\times\pi_2) = E_1^{(i)}(\pi_1)\otimes E_2^{(i)}(\pi_2)$$

and

$$E_{\pm}^{(i)}(\tau(\pi_1\times\pi_2)) = \pm\Theta(E_1^{(i)}(\pi_1)\otimes E_2^{(i)}(\pi_2))$$

where Θ is the mapping defined by

$$\Theta((A_{k\ell})\otimes(B_{rs})) = (A_{r\ell}B_{ks});$$

$E_{\pm}^{(i)}$ is irreducible of dimension d_i^2, and the different $E_{\pm}^{(i)}$ are inequivalent to each other and to $E^{(jk)}$. Since

$$\sum_{i<j}(2d_i d_j)^2 + 2\sum_i d_i^4 = 2(\sum_i d_i^2)^2 = 2|\Pi(Z_1)|^2 = |\hat{\Pi}(Z)|$$

is the order of $\hat{\Pi}(Z)$, there are no other irreducible representations of $\hat{\Pi}(Z)$. The character of $E^{(ij)}$ is found to be

$$\chi^{(ij)}(\pi_1\times\pi_2) = \chi^{(i)}(\pi_1)\chi^{(j)}(\pi_2') + \chi^{(j)}(\pi_1)\chi^{(i)}(\pi_2')$$

and

$$\chi^{(ij)}(\tau(\pi_1\times\pi_2)) = 0;$$

hence by (11.5) the projection is

$$P^{(ij)} = \frac{2d_i d_j}{|\hat{\Pi}(Z)|}\sum_{\pi_1\times\pi_2\in\Pi(Z)}\chi^{(ij)}(\pi_1\times\pi_2)^* I_o \otimes U_1(\pi_1)\otimes U_2(\pi_2)$$

$$= I_o \otimes P_1^{(i)} \otimes P_2^{(j)} + I_o \otimes P_1^{(j)} \otimes P_2^{(i)} .$$

For $E_{\pm}^{(i)}$ we get

$$\chi_{\pm}^{(i)}(\pi_1 \times \pi_2) = \chi^{(i)}(\pi_1)\chi^{(i)}(\pi_2')$$

and

$$\chi_{\pm}^{(i)}(\tau(\pi_1 \times \pi_2)) = \pm \chi^{(i)}(\pi_1 \pi_2')$$

from the definition of the mapping Θ. By (11.5) and (11.10) the projection is

$$P_{\pm}^{(i)} = \frac{1}{2} I_o \otimes P_1^{(i)} \otimes P_2^{(i)} \pm \frac{1}{2} U_o(\tau) \otimes R_i$$

where R_i in $L_2(Z_1) \hat{\otimes} L_2(Z_2)$ is given by

$$R_i f = \frac{2d_i^{\,2}}{|\Pi(Z)|} \sum_{\pi_1 \times \pi_2 \in \Pi(Z)} \chi^{(i)}(\pi_1 \pi_2')^* J^{-1} U_2(\pi_2) f_2 \otimes J U_1(\pi_1) f_1$$

for $f = f_1 \otimes f_2$. Now

$$U_2(\pi_2) = J U_1(\pi_2') J^{-1} = J U_1(\pi_1^{-1}) U_1(\pi_1 \pi_2') J^{-1}.$$

Summing over $\pi_2 \in \Pi(Z_2)$ we get

$$R_i f = \frac{d_i}{|\Pi(Z_1)|} \sum_{\pi_1 \in \Pi(Z_1)} U_1(\pi_1^{-1}) P_1^{(i)} J^{-1} f_2 \otimes J U_1(\pi_1) f_1 .$$

Let $u_1, \ldots, u_{d_j}$ be the orthonormal basis of a $E_1^{(j)}$-generating subspace of $L_2(Z_1)$ such that

$$U_1(\pi_1) u_k = \sum_{r=1}^{d_j} E_1^{(j)}(\pi_1)_{rk} u_r$$

for all k and $\pi_1 \in \Pi(Z_1)$; similarly let

$$U_1(\pi_1) v_h = \sum_{s=1}^{d_m} E_1^{(m)}(\pi_1)_{sh} v_s .$$

Put $f_1=u_k$ and $f_2=Jv_h$; then

$$P_1^{(i)}J^{-1}f_2 = P_1^{(i)}v_h = \delta_{im}v_h$$

and hence by (7.0")

$$R_i(u_k \otimes J_{v_h})$$

$$= \sum_{r=1}^{d_j} \sum_{s=1}^{d_m} \frac{d_i}{|\Pi(Z_1)|} \sum_{\pi_1\in\Pi(Z_1)} E_1^{(m)}(\pi_1)^*_{hs} E_1^{(j)}(\pi_1)_{rk}\delta_{im}v_s\otimes Ju_r$$

$$= \delta_{im}\delta_{jm}v_k\otimes Ju_h.$$

Hence R_i vanishes for elements orthogonal to $P_1^{(i)}L_2(Z_1)\hat{\otimes}P_2^{(i)}L_2(Z_2)$ and can consequently be put in the form $R_i=Q^{(i)}(P_1^{(i)}\otimes P_2^{(i)})$, where $Q^{(i)}(u_{jk}\otimes Ju_{rs})=u_{js}\otimes Ju_{rk}$ follows from our computation and completely determines $Q^{(i)}$. This completes the proof of the lemma.

For the discussion of the exceptional case we need the following concept: An eigenvalue λ of $\overline{S_o(Z_1,D_1,E_1)}$ is called minimal degenerate, if the corresponding eigenspace is a $D_1\otimes E_1$-generating subspace of $L_2(Z_1)$.

11.12 Lemma. Let D_1 and E_1 be irreducible representations of Γ and $\Pi(Z_1)$ respectively such that $P_{D_1}P_{E_1}L_2(Z_1) \neq 0$. For every $\varepsilon > 0$ there exist two orthogonal $D_1\otimes E_1$-generating subspaces N_j of $C_o^\infty(Z_1)$ such that

$$\|S_o(Z_1)u_j-\eta(Z_1,D_1,E_1)u_j\| \leq \varepsilon\|u_j\|$$

for all $u_j\in N_j$ and $j=1,2$ if and only if $\eta(Z_1,D_1,E_1)$ is not an isolated eigenvalue of $\overline{S_o(Z_1,D_1,E_1)}$ which is minimal degenerate.

Proof. Put $A=S_o(Z_1,D_1,E_1)$, $H=P_{D_1}P_{E_1}L_2(Z_1)$ and $\lambda_o=\eta(Z_1,D_1,E_1)$; hence $\lambda_o \in \sigma(A)$. Let $F(.)$ be the spectral family of $\overline{A}$. If λ_o is an isolated eigenvalue of $\overline{A}$ which is minimal degenerate, then for sufficiently small $\varepsilon > 0$, $N=[F(\lambda_o+\varepsilon)-F(\lambda_o-\varepsilon)]H$ is the eigenspace corresponding to λ_o, which by assumption is $D_1 \otimes E_1$-generating. Every such subspace has dimension d equal to the dimension of the representation $D_1 \otimes E_1$. If two subspaces N_1, N_2 with the above properties exist, the dimension of N must be at least $2d=\dim N_1 \oplus N_2$, a contradiction. Now let λ_o not be an isolated eigenvalue of minimal degeneracy; then either $\lambda_o \in \sigma_e(A)$ or λ_o is an eigenvalue of $\overline{A}$ of multiplicity greater than d. By theorems 1.1 and 1.2 it follows that for every $\varepsilon' > 0$ we have

$$\dim\ [F(\lambda_o+\varepsilon')-F(\lambda_o-\varepsilon')]H > d.$$

By theorem 8.11 there exists a $D_1 \otimes E_1$-generating subspace N_1 of D(A) such that

$$\|Au-\lambda_o u\| \leq \varepsilon' \|u\| \quad \text{for every } u \in N_1.$$

Since $\dim N_1=d$, there exists an element $v \neq 0$ of $[F(\lambda_o+\varepsilon')-F(\lambda_o-\varepsilon')]H$ orthogonal to N_1; by approximation there exists an element $w \neq 0$ of D(A) orthogonal to N_1 and such that

$$\|Aw-\lambda_o w\| \leq 2\varepsilon' \|w\|.$$

By part c) of the proof of theorem 8.11, there exists a $D_1 \otimes E_1$-generating subspace N_2 of D(A) orthogonal to N_1 such that

$$\|Au-\lambda_o u\| \leq 2\varepsilon' d^{1/2} \|u\| \quad \text{for all } u \in N_2.$$

Replacing $2\varepsilon' d^{1/2}$ by ε we get the result.

Define

$$(11.13) \quad \eta'(Z_1,D_1,E_1)=\inf\ \{\lambda \in \sigma(S_o(Z_1,D_1,E_1)) : \lambda > \eta(Z_1,D_1,E_1)\}.$$

This number is $>\eta(Z_1,D_1,E_1)$ if and only if $\eta(Z_1,D_1,E_1)$ is an isolated eigenvalue of $\overline{S_o(Z_1,D_1,E_1)}$.

For representations $D\otimes E$ of $\Gamma\times\Pi$ such that $P_D P_E L_2(\mathbb{R}^{3N-3})\neq 0$ and for decompositions $Z=(Z_1,Z_2)$ we now define

$$(11.14)\quad \begin{cases} \eta(Z,D,E)=\inf\ \{\eta(Z_1,D_1,E_1)+\eta(Z_2,D_2,E_2)\} \\ \text{taken over all representations satisfying: } P_{D_i}P_{E_i}\neq 0, \\ \text{there exists } D_o \text{ such that } (D_o,D_1,D_2)<D \text{ and} \\ P_{D_o}L_2(\mathbb{R}^3)\neq 0,\ (E_1,E_2)\underset{Z}{\models\!<} E. \end{cases}$$

However we have two exceptions: 1. If $|Z_2|=1$, then

$$(11.14')\quad \begin{cases} \eta(Z,D,E)=\inf\ \{\eta(Z_1,D_1,E_1)\} \\ \text{taken over all representations satisfying: } P_{D_1}P_{E_1}\neq 0, \\ \text{there exists } D_o \text{ such that } (D_o,D_1)<D \text{ and } P_{D_o}L_2(\mathbb{R}^3)\neq 0, \\ E_1\underset{Z}{|\!<} E \text{ (i.e. the restriction of E to } \Pi(Z_1) \text{ contains } E_1). \end{cases}$$

2. In the exceptional case $|Z_1|=|Z_2|=\frac{N}{2}$ and Z_1,Z_2 identical, replace $\eta(Z_1,D_1,E_1)$ by $\eta'(Z_1,D_1,E_1)$ if the following is true:

$$(11.15)\quad \begin{cases} \text{(i)} & D_1=D_2,\ E_1=E_1^{(i)},\ E_2=E_2^{(i)} \text{ for some } i\in\{1,\dots,p\} \\ & \text{(notation of lemma 11.11).} \\ \text{(ii)} & \eta(Z_1,D_1,E_1) \text{ is an isolated minimal degenerate} \\ & \text{eigenvalue of } \overline{S_o(Z_1,D_1,E_1)}. \\ \text{(iii)} & \text{For all representations } D_o \text{ of } \Gamma \text{ such that} \\ & (D_o,D_1,D_2)<D \text{ and } P_{D_o}L_2(\mathbb{R}^3)\neq 0 \text{ and for all} \\ & \text{representations } E_\omega^{(i)} \text{ of } \hat{\Pi}(Z)\ (\omega=\pm 1) \text{ such} \\ & \text{that } E\,|\,\hat{\Pi}(Z) \text{ contains } E_\omega^{(i)}, \text{ we have } P_\omega P_{D_o}=0, \\ & \text{where } (P_\omega f)(z)=\frac{1}{2}f(z)+\frac{\omega}{2}f(-z) \text{ for } z\in\mathbb{R}^3, \\ & f\in L_2(\mathbb{R}^3) \text{ and } \omega=\pm 1. \end{cases}$$

11.16 Theorem. Let $D\otimes E$ be an irreducible representation of the symmetry group $\Gamma\times\Pi$ such that $P_D P_E \neq 0$. Then the reduced internal Hamiltonian $S_o(D,E)$ has essential spectrum

$$\sigma_e(S_o(D,E)) = [\eta(D,E),\infty[$$

where $\eta(D,E)=\min\{\eta(Z,D,E): Z=(Z_1,Z_2)\}$.

Remark. For an arbitrary irreducible representation $D\otimes E$ of $\Gamma\times\Pi$ with $P_D P_E \neq 0$ we have

$$\eta(D,E) \leq 0, \quad \text{i.e.} \quad [o,\infty[\subset \sigma_e(S_o(D,E)).$$

This is clear for N=2 and follows by induction for the general case.

The proof of theorem 11.16 will be given in section 12. We finish this section with an example in which the exceptional case (11.15) cannot be neglected.

11.17 Example. Let N=4; assume that particles 1 and 3 are identical, the same for 2 and 4, but 1 and 2 are not. The group Π therefore has four elements:

$$\pi_o=1, \pi_1=\begin{pmatrix}1&2&3&4\\3&2&1&4\end{pmatrix}, \quad \pi_2=\begin{pmatrix}1&2&3&4\\1&4&3&2\end{pmatrix}, \quad \pi_3=\begin{pmatrix}1&2&3&4\\3&4&1&2\end{pmatrix}.$$

The Hamiltonian is assumed to have the form

$$S=T+W_{12}+W_{14}+W_{23}+W_{34}$$

where the W_{ij} are spherically symmetric square - well potentials in the internal coordinate x_i-x_j, describing attraction between non-identical particles. The rotation symmetry group is $\Gamma=\mathcal{O}(3)$. The internal Hamiltonians $\hat{S}_{12}, \hat{S}_{14}, \hat{S}_{23}$ and $\hat{S}_{34}$ have identical generating operators which are assumed to have one negative eigenvalue λ_o and no others. This eigenvalue is known to be simple and to have a spherically symmetric eigenfunction. $\hat{S}_{13}$ and $\hat{S}_{24}$ have no eigenvalues.

We choose the representation E of Π given by

$$E(\pi_o)=1,\ E(\pi_1)=-1,\ E(\pi_2)=-1,\ E(\pi_3)=1.$$

The representation D of Γ is fixed later. Let Z be a decomposition with $|Z_2|=1$, say $Z_1=\{1,2,3\}$, $Z_2=\{4\}$. The group $\Pi(Z_1)$ consists of π_o and π_1; $E_1=E|\Pi(Z_1)$ is irreducible and given by $E_1(\pi_o)=1$, $E_1(\pi_1)=-1$. In coordinates $z_1=x_1-x_2$, $z_2=x_3-x_2$ the internal Hamiltonian of Z_1 is

$$S_o(Z_1)=-\frac{1}{2\nu}\Delta_1-\frac{1}{2\nu}\Delta_2-\frac{1}{\mu_2}\,\mathrm{grad}_1\,\mathrm{grad}_2+W_{12}(|z_1|)+W_{23}(|z_2|)$$

where $\frac{1}{\nu}=\frac{1}{\mu_1}+\frac{1}{\mu_2}$. If $\frac{1}{\mu_2}$ is replaced by zero, $S_o(Z_1)$ has one negative eigenvalue, $2\lambda_o$, and no other eigenvalues; the essential spectrum is $[\lambda_o,\infty[$. The eigenvalue $2\lambda_o$ is simple and has an eigenfunction symmetric in z_1, z_2. Since P_{E_1} maps onto the subspace of antisymmetric eigenfuctions, the reduced operator $S_o(Z_1,E_1)$ has $\eta(Z_1,E_1)=\lambda_o$ as lowest point of the spectrum. However

$$-\Delta_1-2\,\mathrm{grad}_1\,\mathrm{grad}_2-\Delta_2 \quad \text{is non-negative;}$$

hence as $\frac{1}{\mu_2}$ increases, the operators $S_o(Z_1)$ and $S_o(Z_1,E_1)$ increase, too. Consequently $\eta(Z_1,D_1,E_1)\geq\lambda_o$ for any D_1, and therefore $\eta(Z,D,E)\geq\lambda_o$ by (11.14') for any D. The same result holds for <u>any</u> decomposition Z such that $|Z_2|=1$.

Now take $|Z_2|=2$; if $Z_1=\{1,3\}$, $Z_2=\{2,4\}$ then $\eta(Z,D,E)=0$ for any D. If $Z_1=\{1,2\}$, $Z_2=\{3,4\}$, the two subsystems are identical; this is the exceptional case. Here $\Pi(Z)=\{\pi_o\}$ and $\hat{\Pi}(Z)=\{\pi_o,\pi_3\}$; hence there is only one possibility: $E_1=1$ and $E_2=1$. If $D_1\neq D_2$, then at least one of the operators $S_o(Z_i,D_i,E_i)$ has $\eta(Z_i,D_i,E_i)=0$ and therefore $\eta(Z,D,E)\geq\lambda_o$. It follows that we only have to consider

$$D_1=D_2=1, \quad \text{where} \quad \eta(Z_i,D_i,E_i)=\lambda_o \quad \text{for } i=1,2$$

and where λ_o is an isolated minimal degenerate eigenvalue, i.e. (11.15.i) and (11.15.ii) hold. Since $E|\Pi(Z)=1$, we have $\omega=1$ in (11.15. iii). If we choose D=1, Then $D_o=1$ is the only possibility, P_{D_o} projects onto the spherically symmetric functions in $L_2(\mathbb{R}^3)$, and hence

$$P_\omega P_{D_o}=P_{D_o}\neq 0.$$

In this case (11.15.iii) does not hold and we get $\eta(Z,D,E)=2\lambda_o$, hence $\eta(D,E)=2\lambda_o$ by theorem 11.16. If however D, and hence D_o, is the representation corresponding to spherical harmonics of order 1, (11.15) holds; it follows that $\eta(Z,D,E)=\lambda_o$ and consequently $\eta(D,E)=\lambda_o$ by theorem 11.16.

12. Proof of theorem 11.16

1. In the first part of the proof we show that

$$[\eta(D,E),\infty[\subset \sigma_e(S_o(D,E));$$

for this it suffices to prove

$$[\eta(Z,D,E),\infty[\ \subset \sigma(S_o(D,E)) \text{ for every } Z = (Z_1,Z_2).$$

a) We start with the case where $|Z_2| > 1$ and Z is not exceptional. By definition (11.14) of $\eta(Z,D,E)$ we have to show that for all representations $(D_o,D_1,D_2) \prec D$ and $(E_1,E_2) \underset{Z}{\models\!<} E$ such that $P_{D_i}P_{E_i} \neq 0$ and $P_{D_o} \neq 0$, and for all $\lambda \geq \eta(Z_1,D_1,E_1) + \eta(Z_2,D_2,E_2)$, we have $\lambda \in \sigma(S(D,E))$. Put

$$\lambda_i = \eta(Z_i,D_i,E_i) \text{ for } i=1,2 \text{ and } \lambda_o = \lambda - \lambda_1 - \lambda_2.$$

Let $\varepsilon > 0$ be given; by theorem 8.11 there exist $D_i \otimes E_i$ - generating subspaces N_i of $C_o^\infty(Z_i)$ such that

$$\|S_o(Z_i)u_i - \lambda_i u_i\| \leq \varepsilon \|u_i\|$$

for all $u_i \in N_i$ and for i=1,2. Since $\lambda_o \geq 0$, we can apply theorem 8.8: for any $r > 0$ there exists a D_o - generating subspace N_o of $C_o^\infty(R^3)$ such that

$$\|T_o u - \lambda_o u\| \leq \varepsilon \|u\|$$

and $u(x) = 0$ for $|x| \leq r$ and for all $u \in N_o$. Define $M = N_o \otimes N_1 \otimes N_2$; then $M \subset C_o^\infty(R^{3N-3})$, and from (11.2) it follows that

$$\|S_o(Z)u - \lambda u\| \leq c_1 \varepsilon \|u\| \text{ for all } u \in M$$

where c_1 depends only on the dimension of M and hence only on the dimension of $D_i \otimes E_i$ and of D_o. Let $\rho > 0$ be such that the support of any $u_i \in N_i$ is contained in the closed ball of radius ρ and center 0 in internal coordinates of Z_i for i=1,2. Let us now transform back to the original coordinates x_j : By lemma 6.8 there exists $r_o > 0$ such that

$$\sum_{j,k \in Z_i} |x_j - x_k|^2 \leq r_o^2 \text{ for } i=1,2$$

and for every x in the support of any function $u \in M$. By definition (11.4) of y_1 we also have

$$|y_1| = \left| \frac{1}{\nu_1} \sum_{j=1}^{n} \mu_j x_j - \frac{1}{\nu_2} \sum_{k=n+1}^{N} \mu_k x_k \right| \geq r$$

for every x in the support of u. It follows that for $s \in Z_1$, $t \in Z_2$ we have

$$|x_s - x_t| = \left| y_1 + \frac{1}{\nu_1} \sum_{j=1}^{n} \mu_j (x_s - x_j) - \frac{1}{\nu_2} \sum_{k=n+1}^{N} \mu_k (x_t - x_k) \right| \geq r - 2r_o .$$

Choosing $r > 3r_o$ and large enough, it follows from (6.7.b) that

$$\|S_o u - S_o(Z)u\| \leq \varepsilon \, (\|u\| + \|\hat{T}_o u\|)$$

for all $u \in M$, where $\hat{T}_o$ is the generating operator in $L_2(\mathbb{R}^{3N-3})$ of the operator $\hat{T}$ of internal kinetic energy. By (6.12), $\hat{T}_o$ is $S_o(Z)$-bounded; hence there is a constant c_2 such that

$$\begin{aligned} \|\hat{T}_o u\| &\leq c_2(\|u\| + \|S_o(Z)u\|) \\ &\leq c_2(1 + |\lambda| + c_1\varepsilon)\|u\| \leq c_3\|u\| \end{aligned}$$

and finally

$$\|S_o u - \lambda u\| \leq \varepsilon(c_1 + 1 + c_3)\|u\| = \varepsilon c_3'\|u\|.$$

for all $u \in M$.

We now construct a non-trivial subspace of $P_D P_E C_o^\infty(\mathbb{R}^{3N-3})$ on which $S_o - \lambda I$ is small. The relation $(D_o, D_1, D_2) \prec D$ implies, by lemma 7.3, that $L = P_D M \neq \{0\}$ and $L \subset M$. As in the proof of theorem 10.5, we show that there is a constant $c_4 > 0$ such that $\|P_E w\| \geq c_4\|w\|$ for all $w \in M$, which implies $K = P_E L = P_D P_E M \neq \{0\}$ and

$$\|(S_o - \lambda I)v\| = \|P_E(S_o - \lambda I)w\| \leq \varepsilon c_3'\|w\| \leq \varepsilon c_3' c_4^{-1}\|v\| \quad \text{for } v = P_E w \in P_E L.$$

Let $\pi \in \Pi \setminus \Pi(Z)$; since Z is not exceptional, we have $\pi Z_1 \cap Z_2 \neq \emptyset$ but $\pi Z_1 \neq Z_2$. Take $j,k \in Z_1$ such that $j \neq k$, $j' = \pi^{-1}(j) \in Z_1$ and $k' = \pi^{-1}(k) \in Z_2$. For $w \in M$ and for x in the support of w we have

$$|x_j - x_k| \leq r_o \text{ and } |x_{j'} - x_{k'}| \geq r - 2r_o > r_o.$$

By (11.3) however, $|x_{j'} - x_{k'}| \leq r_o$ in the support of $U(\pi)w$; hence w and $U(\pi)w$ have disjoint support, and consequently $\langle w|U(\pi)w\rangle = 0$. It follows from definition (7.7) that $\langle w|P_E w\rangle = \langle w|Q_E w\rangle$ where

$$Q_E = \frac{d_E}{|\Pi|} \sum_{\pi \in \Pi(Z)} \chi_E(\pi)^* U(\pi).$$

As in the proof of theorem 10.5, it now follows that there is a number $\nu > 0$ such that $\langle w|Q_E w\rangle = \nu \|w\|^2$, hence $\|P_E w\| \geq \nu^{\frac{1}{2}}\|w\|$ for all $w \in M$.

b) Now let Z be an exceptional decomposition. We try to reason exactly as in part a). If $D_1 \neq D_2$ or if E_1 and E_2 are not equivalent under the group - isomorphism $\pi \mapsto \tau\pi\tau$, then JN_1 is orthogonal to N_2 for any $D_i \otimes E_i$ - generating subspaces N_i of $L_2(Z_i)$, i=1,2, where J is the isometric isomorphism of $L_2(Z_1)$ onto $L_2(Z_2)$ given by (11,9'). If $D_1 = D_2$, and E_1 and E_2 are equivalent, but $\eta(Z_1,D_1,E_1)$ is not an isolated minimal degenerate eigenvalue of $\overline{S_o(Z_1,D_1,E_1)}$, then by lemma 11.12 we can choose N_1 and N_2 such that N_1 and $J^{-1}N_2$ are orthogonal $D_1 \otimes E_1$-generating subspaces of $C_o^\infty(Z_1)$ such that

$$\|S_o(Z_1)u_j - \eta(Z_1,D_1,E_1)u_j\| \leq \varepsilon\|u_j\|$$

for all $u_1 \in N_1$ and $u_2 \in J^{-1}N_2$. It follows that N_2 is $D_2 \otimes E_2$- generating, orthogonal to JN_1, and by (11.9'') we have

$$\|S_o(Z_2)u - \eta(Z_2,D_2,E_2)u\| \leq \varepsilon\|u\| \text{ for all } u \in N_2.$$

The rest of the proof now goes through; in particular we have $\langle w|U(\pi)w\rangle = 0$ for all $w \in M = N_o \otimes N_1 \otimes N_2$ and all $\pi \in \Pi\setminus\Pi(Z)$. This is proved as in a) for $\pi \in \Pi\setminus\hat{\Pi}(Z)$, whereas for $\pi \in \hat{\Pi}\setminus\Pi(Z) = \tau\Pi(Z)$ we have

$$U(\pi)w \in U_o(\tau)N_o \otimes J^{-1}N_2 \otimes JN_1$$

by (11.10), and this space is orthogonal to M.

Next assume that both (11.15.i) and (11.15.ii) hold. Choose N_1 as before and let $N_2 = JN_1$. Again the proof goes through up to the point where we have to consider $\langle w|P_E w\rangle$ for $w \in M$. Since $\langle w|U(\pi)w\rangle = 0$ for $\pi \in \Pi\backslash\hat{\Pi}(Z)$ as before we get $\langle w|P_E w\rangle = \langle w|\hat{Q}_E w\rangle$ where

$$\hat{Q}_E = \frac{d_E}{|\Pi|} \sum_{\pi\in\hat{\Pi}(Z)} \chi_E(\pi)^* U(\pi).$$

The representation $E|\hat{\Pi}(Z)$ is unitarily equivalent to an orthogonal sum of irreducible representations of $\hat{\Pi}(Z)$; hence $\hat{Q}_E$ is a linear combination of the corresponding projections. By lemma 11.11 this gives

$$\hat{Q}_E = \sum_{j<k} a_{jk} P^{(jk)} + \sum_j b_j P_+^{(j)} + \sum_j c_j P_-^{(j)}$$

with non-negative coefficients which are not all zero. By assumption (11.15.i) we have $E_1 = E_1^{(i)}$, $E_2 = E_2^{(i)}$ for some $i \in \{1,\dots,p\}$; since N_1 is $D_1 \otimes E_1$ - generating we have $P_1^{(j)} N_1 = \delta_{ij} N_1$ and similarly $P_2^{(j)} N_2 = \delta_{ij} N_2$ for $j=1,\dots,p$ in the notation of lemma 11.11, hence $P^{(jk)} M = \{0\}$ for all $j < k$ and $P_\pm^{(j)} M = \{0\}$ for $j \neq i$. It follows that

$$\hat{Q}_E w = b_i P_+^{(i)} w + c_i P_-^{(i)} w \quad \text{for } w \in M.$$

If $b_i = c_i = 0$, then $E|\hat{\Pi}(Z)$ does not contain $E_+^{(i)}$ and $E_-^{(i)}$; in this case (11.15.iii) is satisfied. Assume $b_i + c_i > 0$; let $u_1,\dots,u_{d_i}$ be an orthonormal basis of N_1 such that

$$U_1(\pi) u_j = \sum_{r=1}^{d_i} E_1^{(i)}(\pi)_{rj} u_r \quad \text{for all } j \text{ and } \pi \in \Pi(Z_1).$$

Then $Ju_1,\dots,Ju_{d_i}$ is a basis of $N_2 = JN_1$ and we get

$$Q^{(i)}(u_j \otimes Ju_k) = u_j \otimes Ju_k$$

in lemma 11.11; hence $Q^{(i)}$ reduces to the identity on $N_1 \otimes N_2$, and it follows that $P_\pm^{(i)}$ reduces to

$$\left(\tfrac{1}{2} I_o \pm \tfrac{1}{2} U_o(\tau)\right) \otimes I_1 \otimes I_2 = P_\pm \otimes I_1 \otimes I_2$$

on M and consequently

$$P_\pm^{(i)} M = P_\pm N_o \otimes N_1 \otimes N_2.$$

It follows that

$$\langle w|P_E w\rangle = \langle w|\hat{Q}_E w\rangle = b_i\|P_+^{(i)}w\|^2 + c_i\|P_-^{(i)}w\|^2$$

for all $w \in M$.

If $b_i > 0$ and $P_+P_{D_o} \neq 0$, then by corollary 8.8' we can choose N_o such that $P_+N_o = N_o$; hence $\langle w|P_E w\rangle \geq b_i\|w\|^2$ for all $w \in M$. Similarly, if $c_i > 0$ and $P_-P_{D_o} \neq 0$, then we can choose N_o such that $P_-N_o = N_o$ and $\langle w|P_E w\rangle \geq c_i\|w\|^2$ for all $w \in M$. In these cases the proof goes through as before.

The remaining two cases are: (i) $c_i = 0$ (hence $b_i > 0$) and $P_+P_{D_o} = 0$, (ii) $b_i = 0$ (hence $c_i > 0$) and $P_-P_{D_o} = 0$. In these two cases (11.15,iii) is satisfied.

Finally let (11.15) hold. Then we have to show $\lambda \in \sigma(S_o(D,E))$ for all $\lambda \geq \eta'(Z_1,D_1,E_1) + \eta(Z_2,D_2,E_2)$. Let $\lambda_1 = \eta(Z_1,D_1,E_1)$, $\lambda_2 = \eta'(Z_1,D_1,E_1)$ and $\lambda_o = \lambda - \lambda_1 - \lambda_2$. By (11.13) and (11.15.ii) we have $\lambda_2 > \lambda_1$; hence there exist orthogonal $D_1 \otimes E_1$ - generating subspaces L_1, L_2 of $C_o^\infty(Z_1)$ such that

$$\|S_o(Z_1)u_j - \lambda_j u_j\| \leq \varepsilon\|u_j\| \text{ for all } u_j \in L_j \text{ and for } j=1,2.$$

Put $N_1 = L_1$ and $N_2 = JL_2$, and continue the proof as before; this is possible, because JN_1 is orthogonal to N_2.

c) In the case $|Z_2| = 1$ we have to show, according to (11.14'), that $\lambda \in \sigma(S_o(D,E))$ for all $\lambda \geq \eta(Z_1,D_1,E_1)$ and for all D_1, E_1 satisfying the conditions stated in (11.14'). There is no difficulty at all; we omit the details.

2. In the second part of the proof we have to show that $\sigma_e(S_o(D,E)) \subset [\eta(D,E),\infty[$. This is done with the help of theorem 9.4. We first treat the case where Γ and Π are trivial.

a) Consider decompositions $Z = (Z_1,\ldots,Z_m)$ such that $|Z_1| \geq \ldots \geq |Z_m| \geq 1$ and the corresponding operator $S_o(Z)$ defined in section 11. Put $h = N-m+1$ so that $h=1,2,\ldots,N-1$ are the possible values. Identify Z with all decompositions Z' obtained from Z by rearranging the order of subsets. Let Z^{hk}, $k=1,2,\ldots,k(h)$ be an enumeration of all different decompositions for fixed h. For every pair (h,k) define functions $p_{hk\ell r'}$ on $L_2(R^{3N-3})$ for $\ell = 1,\dots,\ell(h,k)$ and $r'> 0$ in the following way: Let $Z^{hk} = (Z_1,\dots,Z_m)$, where $m=N+1-h$, and consider the functions with values

$$(12.1)\qquad \begin{array}{ll} p(x_i-x_j) - (2^{h-1}-1)r' , & i \in Z_r,\ j \in Z_s,\ r \neq s \\ 2^{h-1}r' - p(x_i-x_j) , & i,j \in Z_r \end{array}$$

for $x = (x_1,\dots,x_N) \in R^{3N}$, where $p(z) = (1+|z|^2)^{\frac{1}{2}}$. Restrict to the subspace defined by $\sum_{j=1}^{N} \mu_j x_j = 0$, transform to internal coordinates $y \in R^{3N-3}$, and enumerate the functions in an arbitrary order. If the sets $\Omega(h,k,r',r) \subset R^{3N-3}$ are defined by (9.0), then (9.1) and (9.2) hold. This is proved in lemma 12.6 below. Hence we can employ theorem 9.4.

In theorem 9.4 replace m by 3N-3, S by S_o, T by $\hat{T}_o$, the generating operator of $\hat{T}$ (It has been shown in section 6, that $\hat{T}_o = -\Delta$ in suitable internal coordinates). Assumption (i) of theorem 9.4 is satisfied because of (6.7.c). In order to satisfy (ii), we have to choose V_{hk}. Note that by (12.1) and (9.0) we have $p(x_i-x_j) > s$ for all x=ay such that $y \in \Omega(h,k,r',s)$, and for all i,j belonging to different subsets of Z^{hk}. By (6.7.b) an operator $W_{j_1,\dots,j_n}$ satisfies the assumption for V_{hk}, if $\{j_1,\dots,j_n\}$ intersects at least two subsets of Z^{hk}. It follows that we can take $V_{hk} = S_o - S_o(Z)$, where Z is any decomposition into two subsets Z_1,Z_2 such that every subset of Z^{hk} is contained either in Z_1 or in Z_2. Then $S_{hk} = S_o(Z)$,

$$\eta_{hk} = \min\sigma(S_{hk}) = \min\sigma(S_o(Z_1)) + \min\sigma(S_o(Z_2)) = \eta(Z_1) + \eta(Z_2)$$

by (11.2) for $|Z_2| > 1$ and $\eta_{hk} = \eta(Z_1)$ by (11.2') for $|Z_2|=1$. Hence, by theorem 9.4, $\sigma_e(S_o) \subset [\eta,\infty[$, where $\eta = \min\{\eta_{hk}:h,k\}$ coincides with $\eta(D,E)$ in the special case which we consider here.

b) To treat the general case, we have to modify theorem 9.4 as we did in the proof of theorem 10.5. Replace $C_o^\infty(R^{3N-3})$ by

$$D(S_o(D,E)) = P_D P_E C_o^\infty(R^{3N-3}),$$

assuming $P_D P_E \neq 0$. By (12.1) all functions $p_{hk\ell r'}$ are invariant under Γ; hence the same is true for the functions Φ_{hks} defined by (9.5). For $\pi \in \Pi$ the functions $p_{hk\ell r'}$ are interchanged by the substitution $x \mapsto \pi^{-1}x$ in such a way, that h is kept fixed; if we define $U(\pi)f$ by (11.3) for any function on R^{3N-3}, we get $U(\pi)\,\Phi_{hks} = \Phi_{hk's}$ where k' is such that $Z^{hk'} = \pi Z^{hk}$. It follows that the functions

$$\sum_{k=1}^{k(h)} \Phi_{hks} \text{ are invariant under } \Gamma \text{ and } \Pi;$$

hence for $u \in D(S_o(D,E))$, the functions v_h defined by (9.9) are also in $D(S_o(D,E))$.

Now we consider the special case $h = N-1$ where $Z^{hk} = (Z_1, Z_2)$; suppose that this decomposition is not exceptional. By lemma 7.5 and by (7.13) we have

$$D(S_o(D,E)) \subset \oplus[D(T_o,D_o) \hat{\otimes} D(S_o(Z_1,D_1,E_1)) \hat{\otimes} D(S_o(Z_2,D_2,E_2))]$$

where the sum is taken over all $(D_o,D_1,D_2) \prec D$ and all $(E_1,E_2) \underset{Z^{hk}}{\models\!\!\prec} E$. By (12.1) the functions $p_{hk\ell r'}$ are interchanged by the substitution $x \mapsto \pi^{-1}x$ with $\pi \in \Pi(Z^{hk})$ in such a way, that h and k are kept fixed; hence by (9.5) the functions Φ_{hks} are invariant under $\Pi(Z^{hk})$. It follows from (9.9) that

$$w_{hk} \in \oplus\, [D(T_o,D_o) \hat{\otimes} D(S_o(Z_1,D_1,E_1)) \hat{\otimes} D(S_o(Z_2,D_2,E_2))]$$

and consequently by (11.14) the inequality (9.15) holds with η_{hk} replaced by $\eta(Z^{hk},D,E)$.

Next we assume that $h=N-1$, and Z^{hk} is exceptional. By lemmas 7.5 and 7.9 we have

$$D(S_o(D,E)) \subset \oplus\, P_{\hat{E}}[P_{D_o}C_o^\infty(Z_o) \hat{\otimes} P_{D_1}C_o^\infty(Z_1) \hat{\otimes} P_{D_2}C_o^\infty(Z_2)]$$

where the sum is taken over all $(D_o,D_1,D_2) \prec D$ and all irreducible representations $\hat{E}$ of $\hat{\Pi}(Z^{hk})$ contained in $E|\hat{\Pi}(Z^{hk})$. Again Φ_{hks} is invariant under $\hat{\Pi}(Z^{hk})$ and therefore

$$w_{hk} \in \oplus\, P_{\hat{E}}[P_{D_o}C_o^\infty(Z_o) \hat{\otimes} P_{D_1}C_o^\infty(Z_1) \hat{\otimes} P_{D_2}C_o^\infty(Z_2)].$$

For all components of w_{hk} corresponding to representations

$D_1 \neq D_2$ or $\widehat{E} = E^{(ij)}$ (see lemma 11.11) we get the estimate (9.15) with η_{hk} replaced by $\eta(Z_1,D_1,E_1^{(i)}) + \eta(Z_2,D_2,E_2^{(j)})$. It remains to treat the components corresponding to $D_1 = D_2$ and $\widehat{E} = E_\omega^{(i)}$. If $\eta(Z_1,D_1,E_1^{(i)})$ is not an isolated minimal degenerate eigenvalue of $\overline{S_o(Z_1,D_1,E_1^{(i)})}$ or if $P_\omega P_{D_o} \neq 0$, we obtain the same estimate as before using the explicit formula for $P_\omega^{(i)}$ given by lemma 11.11.

Finally we are left with the case (11.15). Let N_1 be the eigenspace of $\overline{S_o(Z_1,D_1,E_1^{(i)})}$ for the eigenvalue $\eta(Z_1,D_1,E_1^{(i)})$ and let

$$H_1 = N_1^\perp \cap D(\overline{S_o(Z_1,D_1,E_1^{(i)})}).$$

By definition (11.13) of $\eta'(Z_1,D_1,E_1^{(i)})$ this number is the minimum of the spectrum of $\overline{S_o(Z_1,D_1,E_1^{(i)})}$ restricted to H_1. It follows that

$$P_{D_1} P_{E_1^{(i)}} C_o^\infty(Z_1) \subset N_1 \oplus H_1$$

and similarly

$$P_{D_2} P_{E_2^{(i)}} C_o^\infty(Z_2) \subset JN_1 \oplus JH_1 = N_2 \oplus H_2.$$

If w is the component of w_{hk} corresponding to $E_\omega^{(i)}$, D_o, D_1, D_2, then by lemma 11.11

$$w \in P_\omega^{(i)}[P_{D_o} C_o^\infty(Z_o) \hat{\otimes} (N_1 \oplus H_1) \hat{\otimes} (N_2 \oplus H_2)]$$

and we get four orthogonal components. The first component vanishes because it is contained in

$$P_\omega^{(i)}[P_{D_o} C_o^\infty(Z_o) \hat{\otimes} N_1 \hat{\otimes} N_2] = P_\omega P_{D_o} C_o^\infty(Z_o) \hat{\otimes} N_1 \hat{\otimes} N_2 = \{0\}$$

by lemma 11.11 and by (11.15.iii). Hence it remains to estimate

$$\langle w | S_o(Z^{hk}) w \rangle = \langle w | \overline{S_o(Z^{hk})} w \rangle$$

for

$$w \in P_\omega^{(i)}[P_{D_o} C_o^\infty(Z_o) \hat{\otimes} \{(N_1 \hat{\otimes} H_2) \oplus (H_1 \hat{\otimes} N_2) \oplus (H_1 \hat{\otimes} H_2)\}].$$

By (11.2) we have

$$\overline{S_o(Z^{hk})} = \overline{T}_o \circ I_1 \circ I_2 + I_o \circ \overline{S_o(Z_1)} \circ I_2 + I_o \circ I_1 \circ \overline{S_o(Z_2)}.$$

Now the estimate (9.15) follows for w with η_{hk} replaced by $\eta'(Z_1,D_1,E_1^{(i)}) + \eta(Z_2,D_2,E_2^{(i)})$. Putting all estimates together we get (9.15) for w_{hk} with η_{hk} replaced by $\eta(Z^{hk},D,E)$.

For $h < N-1$ let $Z^{hk} = (Z_1,\dots,Z_m)$, $m=N+1-h > 2$ and $|Z_1| \geq |Z_2| \geq \dots \geq |Z_m|$. We introduce internal coordinates for $S(Z^{hk})$ as follows: Denote by m' the number of subsystems Z_j with $|Z_j| > 1$; the extreme case is $m' = 0$ where $|Z_j| = 1$ for all j. Introduce internal coordinates for each of the subsystems Z_j with $j \leq m'$ (if any), denote by $L_2(Z_j)$ the corresponding spaces and by $S_o(Z_j)$ the generating operator for the internal Hamiltonian of the subsystem Z_j. Let $z_1,\dots,z_m$ be the center-of-mass coordinates of the subsystems and $\nu_1,\dots,\nu_m$ their masses. In these coordinates we have

$$\hat{S}(Z^{hk}) = - \sum_{j=1}^{m} \frac{1}{2\nu_j} \Delta_j + \sum_{j=1}^{m'} \hat{S}(Z_j)$$

where Δ_j is the Laplacean with respect to z_j.
Now choose internal coordinates $y_1,\dots,y_{m-1}$ for the first part such that

$$- \sum_{j=1}^{m} \frac{1}{2\nu_j} \Delta_j = - \sum_{j=1}^{m-1} \frac{1}{2\varkappa_j} \Delta_j' + \tilde{T}$$

where Δ_j' is the Laplacean with respect to y_j and where $\tilde{T}$ is the operator of translation energy of the total system given by (6.5'). This is achieved by putting $y_1' = z_1, \varkappa_1' = \nu_1$ and

$$y_j = z_{j+1} - y_j' \quad , \quad \varkappa_j = \left(\frac{1}{\varkappa_j'} + \frac{1}{\nu_{j+1}}\right)^{-1}$$

(12.2)

$$y_{j+1}' = (\varkappa_{j+1}')^{-1}(\varkappa_j' y_j' + \nu_{j+1} z_{j+1}), \quad \varkappa_{j+1}' = \varkappa_j' + \nu_{j+1}$$

for $j=1,\dots,m-1$. Let $L_2(Z_o)$ be the L_2 - space of functions of $y_1,\dots,y_{m-1}$. Then

(12.3) $$L_2(R^{3N-3}) = \hat{\bigotimes}_{j=0,\dots,m'} L_2(Z_j)$$

and

(12.4) $$S_o(Z^{hk}) = \sum_{j=0}^{m'} I_o \otimes \dots \otimes I_{j-1} \otimes S_o(Z_j) \otimes \dots \otimes I_{m'},$$

where

(12.4') $$S_o(Z_o) = - \sum_{j=1}^{m-1} \frac{1}{2\varkappa_j} \Delta_j'$$

and where I_j is the identity in $L_2(Z_j)$.

We have to obtain a lower bound for $\langle w_{hk} | S_o(Z^{hk}) w_{hk} \rangle$. Since Φ_{hks} is invariant under $\Pi(Z^{hk}) = \Pi(Z_o) \times \dots \times \Pi(Z_{m'})$ we get from (9.9)

$$w_{hk} \in \oplus[D(S_o(Z_o,D_o)) \hat{\otimes} D(S_o(Z_1,D_1,E_1) \hat{\otimes} \dots \hat{\otimes} D(S_o(Z_{m'},D_{m'},E_{m'}))]$$

where the sum is taken over all $(D_o,\dots,D_{m'}) \prec D$ and all $(E_1,\dots,E_{m'}) \underset{Z^{hk}}{\models} E$. Since $S_o(Z_o)$ is non-negative, it follows from (12.4) that (9.15) holds with η_{hk} replaced by

(12.5) $$\inf \sum_{j=1}^{m'} \eta(Z_j,D_j,E_j)$$

taken over all representations D_j, E_j such that : $P_{D_j} P_{E_j} \neq 0$, there exists D_o such that $P_{D_o} \neq 0$ and $(D_o,\dots,D_{m'}) \prec D$, $(E_1,\dots,E_{m'}) \underset{Z^{hk}}{\models} E$.

Now write

$$L_2(Z_o) = L_2(Z_{oo}) \hat{\otimes} L_2(\tilde{Z}_o)$$

where $L_2(Z_{oo})$ corresponds to the variable y_1 introduced in (12.2). Then by construction the space

$$L_2(\tilde{Z}_1) = L_2(Z_{oo}) \hat{\otimes} L_2(Z_1) \hat{\otimes} L_2(Z_2)$$

is the space corresponding to internal coordinates for the subsystem $Z_1 \cup Z_2$ and

$$L_2(\tilde{Z}_o) \hat{\otimes} L_2(\tilde{Z}_1) \hat{\otimes} L_2(Z_3) \hat{\otimes} \dots \hat{\otimes} L_2(Z_{m'})$$

is the representation of $L_2(R^{3N-3})$ analogous to (12.3) for the decomposition $\tilde{Z} = (\tilde{Z}_1, Z_3, \dots, Z_m)$. Take one choice of representations $D_o,\dots,D_{m'}$ and $E_1,\dots,E_{m'}$ occuring in (12.5). Let D_{oo} and $\tilde{D}_o$ be such that $(D_{oo},\tilde{D}_o) < D_o$ and $P_{D_{oo}} \neq 0$, $P_{\tilde{D}_o} \neq 0$; then

$$(D_{oo},\tilde{D}_o,D_1,\ldots,D_{m'}) < D$$

and, by applying lemma 7.6 twice, we find $\tilde{D}_1$ such that

$$(D_{oo},D_1,D_2) < \tilde{D}_1 \text{ and } (\tilde{D}_o,\tilde{D}_1,D_3,\ldots,D_{m'}) < D.$$

By lemma 7.4, $\tilde{D}_1$ can be chosen such that $P_{\tilde{D}_1} \neq 0$. Similarly, by lemma 7.12, an irreducible representation $\tilde{E}$ of $\Pi(Z_1 \cup Z_2)$ exists such that

$$P_{\tilde{E}} \neq 0 \text{ and } (E_1,E_2) \underset{Z}{\ltimes} \tilde{E}, \text{ where } Z = (Z_1,Z_2),$$

and such that

$$(\tilde{E},E_3,\ldots,E_{m'}) \underset{\tilde{Z}}{\ltimes} E.$$

By part 1.a) of the proof, we know that if the decomposition (Z_1,Z_2) of $Z_1\cup Z_2$ is not exceptional, then $\eta(Z_1,D_1,E_1) + \eta(Z_2,D_2,E_2)$ is in the spectrum of $S_o(Z_1\cup Z_2,\tilde{D}_1,\tilde{E})$. It follows that

$$\eta(Z_1\cup Z_2,\tilde{D}_1,\tilde{E}) \leq \eta(Z_1,D_1,E_1) + \eta(Z_2,D_2,E_2)$$

and hence the lower bound (12.5) can be replaced by the similar bound for decompositions of $\{1,\ldots,N\}$ into m-1 subsets. Continuing this process, we get the lower bound obtained previously for decompositions into two subsets.

The above reasoning fails in the exceptional case, where a transformation $\tau \in \Pi$ exists which maps Z_1 onto Z_2, and vice versa, and Z_j onto itself for $j > 2$. However, in this case it suffices to replace $\Pi(Z^{hk})$ by

$$\tilde{\Pi}(Z^{hk}) = \hat{\Pi}(Z_1,Z_2)\times \Pi(Z_3) \times\ldots\times \Pi(Z_{m'}),$$

where

$$\hat{\Pi}(Z_1,Z_2) = (\Pi(Z_1)\times \Pi(Z_2)) \cup \tau(\Pi(Z_1)\times \Pi(Z_2)),$$

and to rely on part 1.b) of the proof. We omit the details. It is important to note, however, that the exceptional case can occur only in the first step of the reduction, because $|Z_1\cup Z_2| = |Z_1|+|Z_2|>|Z_3|$.

Putting all estimates together, theorem 9.4 gives the lower bound $\eta(D,E)$ for the essential spectrum of $S_o(D,E)$. This completes the proof.

The following lemma was used in the course of the proof:

12.6 Lemma For the functions $p_{hk\ell r'}$ defined by (12.1) and $\Omega(h,k,r',r)$ defined by (9.0) with these functions, the assertions (9.1) and (9.2) hold.

Proof. We prove the statements in the x coordinates with $\alpha = 2$; transformation to internal coordinates gives the result with a different value for α.

1. (Proof of (9.1.i)). For h=1 there is only one decomposition Z^{11}; hence we may take $h \geq 2$. Let $k \neq k'$, $s > 0$, $t > 0$, $r' > 0$ and $s + t - r' > 0$. There exist i,j with $i \neq j$ such that i and j are contained in one subset of Z^{hk} but in different subsets of $Z^{hk'}$. Hence by (12.1), for $x \in \Omega(h,k,r',s)$ and $y \in \Omega(h,k',r',t)$ it follows that

$$2^{h-1}r' - p(x_i - x_j) > s \text{ and } p(y_i - y_j) - (2^{h-1}-1)r' > t.$$

By (10.7.ii) we get

$$|x-y| \geq \max\{|x_i-y_i|,|x_j-y_j|\} \geq \frac{1}{2}|x_i-x_j-y_i+y_j|$$

$$\geq \frac{1}{2}p(y_i-y_j) - \frac{1}{2}p(x_i-x_j) > \frac{1}{2}(s+t-r'), \quad \text{q.e.d.}$$

2. (Proof of (9.1.ii)). Let $s > 0$, $t > 0$, $r' > 0$, $s-t > 0$, $x \in \Omega(h,k,r',s)$ and $y \in \complement\Omega(h,k,r',t)$. It follows that either there exists a pair i,j such that i and j belong to different subsets of Z^{hk} and such that

$$p(y_i-y_j) - (2^{h-1}-1)r' \leq t$$

or there is a pair i,j with $i \neq j$ belonging to one subset of Z^{hk} and such that

$$2^{h-1}r' - p(y_i-y_j) \leq t.$$

In the first case we have $p(x_i-x_j) - (2^{h-1}-1)r' > s$ and therefore

$$|x-y| \geq \max\{|x_i-y_i|, |x_j-y_j|\} \geq \frac{1}{2} |x_i-y_i-x_j+y_j|$$

$$\geq \frac{1}{2} p(x_i-x_j) - \frac{1}{2} p(y_i-y_j) > \frac{1}{2} (s-t).$$

In the second case we have $2^{h-1}r' - p(x_i-x_j) > s$ and therefore

$$|x-y| > \frac{1}{2} (s-t)$$

by the same reasoning.

3.(Proof of (9.2)). Let $0 < r < r'$ and $x \in \complement(\bigcup_{h,k} \Omega(h,k,r',r))$, hence $x \in \complement\Omega(h,k,r',r)$ for all h,k. For $h=1$ there is only one decomposition into N subsets containing one element each. Hence $x \in \complement\Omega(1.1.r',r)$ implies that there is a pair j_1, j_2 with $j_1 \neq j_2$ such that

$$p(x_{j_1} - x_{j_2}) \leq r.$$

For $h=2$ choose Z^{2k} such that $Z_1 = \{j_1, j_2\}$ is the first subset. Then all other subsets have only one element each. Hence $x \in \complement\Omega(2,k,r',r)$ implies either $2r' - p(x_{j_1} - x_{j_2}) \leq r$ or there exists a pair i,j belonging to different subsets and satisfying $p(x_i-x_j) - r' \leq r$. The first case is not possible, because $2r' - p(x_{j_1} - x_{j_2}) \geq 2r' - r > r$. In the second case either i and j both are different from j_1, j_2, then we get a pair j_3, j_4 with $j_3 \neq j_4$ such that

$$p(x_{j_3} - x_{j_4}) \leq r' + r;$$

or one of the two numbers i, j is in Z_1, then there is a j_3 disjoint from Z_1 such that

$$p(x_{j_1} - x_{j_3}) \leq r' + r \text{ or } p(x_{j_2} - x_{j_3}) \leq r' + r.$$

In any case we get a triple j_1, j_2, j_3 such that

$$p(x_{j_i} - x_{j_k}) \leq r' + 2r \text{ for } i,k=1,2,3$$

or two pairs j_1, j_2 and j_3, j_4 such that

$$p(x_{j_1} - x_{j_2}) \leq r' + 2r \text{ and } p(x_{j_3} - x_{j_4}) \leq r' + 2r.$$

For h=3 we choose Z^{3k} according to the result of the previous section: Either choose $Z_1 = \{j_1, j_2, j_3\}$ or $Z_1 = \{j_1, j_2\}$, $Z_2 = \{j_3, j_4\}$. Reasoning as above we get either a quadruple j_1, j_2, j_3, j_4 or a triple j_1, j_2, j_3 and a pair j_4, j_5 or three pairs j_1, j_2 and j_3, j_4 and j_5, j_6 such that

$$p(x_i - x_j) \leq 4r' + 3r$$

for i,j belonging to one of these subsets.

By induction we get for arbitrary $h \in \{1, \ldots, N-1\}$ either a (h+1)-tuple or a h-tuple and a pair and so on, such that

$$p(x_i - x_j) \leq (2^h - h - 1)r' + hr$$

for i,j belonging to one of these subsets. For h=N-1 there is only one possibility: here we get an N-tuple, hence

$$p(x_i - x_j) \leq (2^{N-1} - N)r' + (N-1)r < 2^{N-1}r' \text{ for } i,j = 1, \ldots, N.$$

Now we have to restrict to the subspace defined by $\sum_{j=1}^{N} \mu_j x_j = 0$. It follows that $x_i = \frac{1}{\mu} \sum_{j=1}^{N} \mu_j (x_i - x_j)$ for every i, hence

$$|x_i| \leq \frac{1}{\mu} \sum_{j=1}^{N} \mu_j |x_i - x_j| \leq \frac{1}{\mu} \sum_{j=1}^{N} \mu_j p(x_i - x_j) \leq 2^{N-1} r'$$

for all i, which shows that $\complement(\bigcup_{h,k} \Omega(h,k,r',r))$ is contained in a compact set K(r') depending on r' only. This completes the proof of the lemma.

Appendix: Extension to N-particle systems with spin

In this final section we shall indicate how our results can be extended to N-particle systems with spin. We assume that the j-th particle can occupy m_j different spin states. Then the state space for the system consisting of the j-th particle only, is

$$L_2(\mathbb{R}^3 \times M_j) = L_2(\mathbb{R}^3) \otimes \mathbb{C}^{m_j},$$

where $M_j = \{1,\dots,m_j\}$ is the domain of the spin coordinate of the j-th particle. The state space of the entire system is then given by

$$\hat{\bigotimes_{j=1}^{N}} L_2(\mathbb{R}^3 \times M_j) = L_2(\mathbb{R}^{3N} \times M) = L_2(\mathbb{R}^{3N}) \otimes \Big(\bigotimes_{j=1}^{N} \mathbb{C}^{m_j}\Big),$$

where $M = \prod_{j=1}^{N} M_j$.

The total Hamiltonian is assumed to have the form

$$S = K + V,$$

where

$$D(S) = C_o^\infty(\mathbb{R}^{3N} \times M) = C_o^\infty(\mathbb{R}^{3N}) \otimes \Big(\bigotimes_{j=1}^{N} \mathbb{C}^{m_j}\Big);$$

K is an operator in $L_2(\mathbb{R}^{3N} \times M)$ acting only on $\mathbb{R}^{3N}$ (the space coordinates of the system), generated by the operator T of kinetic energy, as it was defined in section 6. The operator V has the form

$$V = \sum_{n=1}^{N} \sum_{j_1 < \dots < j_n} V_{j_1 \dots j_n} + \sum_{n=2}^{N} \sum_{j_1 < \dots < j_n} W_{j_1 \dots j_n},$$

where the operators $V_{j_1\dots j_n}$ and $W_{j_1\dots j_n}$ have the same physical meaning as in section 6.

The operator $V_{j_1\dots j_n}$ acts only on the (space and spin) coordinates of the particles $j_1,\dots,j_n$ and is generated by a symmetric operator $V^{j_1\dots j_n}$ in $L_2(\mathbb{R}^{3n}\times(\prod_{i=1}^{n} M_{j_i}))$; $V^{j_1\dots j_n}$ is $K^{j_1\dots j_n}$-small at infinity, where $K^{j_1\dots j_n}$ acts only on $\mathbb{R}^{3n}$ and is generated by the operator $T^{j_1\dots j_n}$ of section 6.

In order to state our conditions on $W_{j_1\dots j_n}$ we have to say something about internal coordinates. Let $\{j_1,\dots,j_n\}$ be any subsystem ($n \geq 2$) of $\{1,2,\dots,N\}$; the center of mass coordinate and the internal space coordinates are defined as in section 6 (not involving spin coordinates). Furthermore the spin coordinates of the subsystem are considered as internal coordinates.

Let now the unitary operator U corresponding to the subsystem $\{j_1,\dots,j_n\}$ be defined as in section 6 (acting only on the space coordinates). Then $W_{j_1\dots j_n}$ is an operator in $L_2(\mathbb{R}^{3N}\times M)$ such that $UW_{j_1\dots j_n}U^*$ acts only on the internal coordinates of the subsystem $\{j_1,\dots,j_n\}$ and is generated by a symmetric operator $W^{j_1\dots j_n}$ in $L_2(\mathbb{R}^{3(n-1)}\times\prod_{j=1}^{n} M_{j_i})$, which is Δ-small at infinity (here Δ is the operator in $L_2(\mathbb{R}^{3(n-1)}\times\prod_{i=1}^{n} M_{j_i})$ which acts only on $\mathbb{R}^{3(n-1)}$ and is generated by the operator Δ in $L_2(\mathbb{R}^{3(n-1)})$).

Finally we have to replace condition (6.12): For any subsystem $\{j_1,\dots,j_n\}$ the total and internal Hamiltonians $S_{j_1\dots j_n}$ and $\hat{S}_{j_1\dots j_n}$ are bounded below; $K_{j_1\dots j_n}$ is $S_{j_1\dots j_n}$-bounded and $\hat{K}_{j_1\dots j_n}$ is $\hat{S}_{j_1\dots j_n}$-bounded, where $K_{j_1\dots j_n}$ and $\hat{K}_{j_1\dots j_n}$ are the total and the internal Hamiltonian of the subsystem respectively.

We have now to explain the symmetry properties of these systems and operators. Permutation symmetry is clear, for $\pi \in \Pi$ the unitary operator $U(\pi)$ is given by the corresponding permutation of the (space and spin) coordinates of the particles. For an irreducible unitary representation E of Π the projection P_E is defined formally as in section 7.

To study the rotation-reflection symmetry let us first consider the system consisting of the j-th particle only. As symmetry group one has to consider the group $\hat{\Gamma}$ which is constructed from the 2-dimensional unitary group by adjoining one abstract element ($\hat{\Gamma}$ is a covering group of $\mathfrak{O}(3)$; if m_j is odd, one can use $\mathfrak{O}(3)$ instead of $\hat{\Gamma}$; see [6] Sect.9-6 and [35] Sect.15). In general the symmetry group (with respect to rotation-reflection symmetry) is some compact subgroup Γ of $\hat{\Gamma}$. There are group homomorphisms d and d_j

$$d:\Gamma \to \mathfrak{O}(3), \quad d_j:\Gamma \to \mathcal{U}(m_j)$$

where $\mathcal{U}(m)$ is the m-dimensional unitary group. Then

$$U^{(j)}(\gamma):=U_{d(\gamma)}\otimes d_j(\gamma)$$

with

$$U_{d(\gamma)}f(x):=f(d(\gamma)^{-1}x)$$

defines a unitary representation of Γ in $L_2(\mathbb{R}^3)\otimes\mathbb{C}^{m_j}$. Similarly

$$U(\gamma):=\bigotimes_{j=1}^{N} U^{(j)}(\gamma)$$

and

$$U(\gamma):=\bigotimes_{j=1}^{N-1} U_{d(\gamma)}\otimes\Big(\bigotimes_{j=1}^{N} d_j(\gamma)\Big)$$

define unitary representations of Γ in $L_2(\mathbb{R}^{3N}\times M)$ and $L_2(\mathbb{R}^{3(N-1)}\times M)$ respectively. The projections P_D are defined by integrating over Γ with respect to the invariant probability measure of Γ.

Now all our theorems can be restated for this new situation and the proofs go through in exactly the same way. It is important to note that the operator of the kinetic energy of the center of mass motion of any system, and the operator of the kinetic energy of the relative motion of two systems are operators in $L_2(\mathbb{R}^3)$ as before. In the formulation of theorem 8.11 $\Gamma \subset \mathfrak{S}(3)$ can be replaced by $\Gamma \subset \hat{\Gamma}$; the proof need not be changed. Theorem 9.4 holds in the same form for operators in $L_2(\mathbb{R}^m \times \mathbb{C}^k)$, if T is replaced by the operator which acts only on $\mathbb{R}^m$ and is generated by $-\Delta$.

References

[1] Albeverio, S.: On bound states in the continuum of N-body systems and the virial theorem (to appear).

[2] Balslev, E.: Spectral Theory of Schrödinger Operators of Many-body Systems with Permutation and Rotation Symmetries (to appear).

[3] Brascamp, H.J. and C. van Winter: The N-body Problem with Spin-Orbit or Coulomb Interactions. Comm. Math. Phys. 11, 19-55 (1968/69).

[4] Combes, J.M.: Relatively Compact Interactions in Many Particle Systems. Comm. Math. Phys. 12, 283-295 (1969).

[5] Goldberg, S.: Unbounded linear operators. McGraw-Hill (1966).

[6] Hamermesh, M.: Group theory and its application to physical problems. Addison-Wesley (1962).

[7] Helgason, S.: Differential operators on homogeneous spaces. Acta Math. 102, 239-299 (1959).

[8] Hunziker, W.: Proof of a Conjecture of S. Weinberg. Phys. Rev. 135 Nr. 3 B, 800-803 (1964).

[9] - : On the Spectra of Schrödinger Multiparticle Hamiltonians. Helvetica phys. acta 39, 451-462 (1966).

[10] Ikebe, T. and T. Kato: Uniqueness of the Self-adjoint Extension of Singular Elliptic Differential Operators. Archive Rat. Mech. Analysis 9, 77-92 (1962).

[11] Jörgens, K.: Wesentliche Selbstadjungiertheit singulärer elliptischer Differentialoperatoren zweiter Ordnung in $C_o^\infty(G)$. Math. Scand. 15, 5-17 (1964).

[12] - : Über das wesentliche Spektrum elliptischer Differentialoperatoren vom Schrödinger-Typ. Tech. Report, Univ. of Heidelberg (1965).

[13] - : Zur Spektraltheorie der Schrödinger-Operatoren. Math. Zeitschr. 96, 355-372 (1967).

[14] - : Spectral Theory of Schrödinger Operators. Lecture notes, University of Colorado (1970).

[15] Kato, T.: Perturbation theory for linear operators. Springer (1966).

[16] - : Fundamental properties of Hamiltonian operators of Schrödinger type. Trans. Amer. Math. Soc. 70, 195-211 (1951).

[17] - : On the existence of solutions of the helium wave equation. Trans. Amer. Math. Soc. 70, 212-219 (1951).

[18] - : Growth properties of solutions of the reduced wave equation with a variable coefficient. Comm. Pure Appl. Math. 12, 403-425 (1959).

[19] - : Some mathematical problems in quantum mechanics. Supplement of the Progr. of Theor. Phys. No. 40 (1967).

[20] Maurin, K.: General eigenfunction expansions and unitary representations of topological groups. Polish Scientific Publishers (1968).

[21] Neumark, M.A.: Lineare Darstellungen der Lorentzgruppe. VEB Deutscher Verlag der Wissenschaften (1963).

[22] Sigalov, A.G.: The main mathematical problem in the theory of atomic spectra. Uspehi mat. Nauk 22 Nr. 2 (134), 3-20 (1967)(russian); Russ.Math. Surveys 22 Nr. 2, 1-18 (1967).

[23] Simon, B.: On the infinitude or finiteness of the number of bound states of an N-body quantum system. Helv. Physica Acta 43, 607-630 (1970).

[24] - : Quantum mechanics for Hamiltonians defined as quadratic forms. Princeton University Press (1971).

[25] Stummel, F.: Singuläre elliptische Differentialoperatoren in Hilbertschen Räumen. Math. Annalen 132, 150-176 (1956).

[26] Uchiyama, J.: On the discrete eigenvalues of the many-particle system. Publ. RIMS, Kyoto Univ. Ser. A 2, No. 5, 117-132 (1966).

[27] - : Finiteness of the number of discrete eigenvalues of Schrödinger operator for a three particle system. Publ. RIMS, Kyoto Univ. Ser. A 5, No. 1, 51-63 (1969).

[28] - : Corrections to "finiteness of the number of discrete eigenvalues of the Schrödinger operator for a three particle system". Publ. RIMS, Kyoto Univ. Ser. A 6, No. 1, 189-192 (1970).

[29] - : Finiteness of the number of discrete eigenvalues of the Schrödinger operator for a three particle system II. Publ. RIMS, Kyoto Univ. Ser. A 6, No. 1, 193-200 (1970).

[30] - : On the existence of the discrete eigenvalue of the Schrödinger operator for the negative hydrogen Ion. Publ.RIMS Kyoto Univ. Ser. A 6, No. 1, 201-204 (1970).

[31] Weidmann, J.: On the continuous spectrum of Schrödinger operators. Comm. Pure Appl. Math. 19, 107-110 (1966).

[32] - : The virial theorem and its application to the spectral theory of Schrödinger operators. Bull. Amer. Math. Soc. 73, 452-456 (1967).

[33] Weinberg, S.: Systematic solution of multiparticle scattering problems. Phys. Rev. 133 B, 232-256 (1964).

[34] Weyl, H.: Gruppentheorie und Quantenmechanik. S. Hirzel 1928.

[35] Wigner, E.P.: Group theory and its application to the quantum mechanics of atomic spectra. Academic Press (1959).

[36] Winter, C. van: Theory of finite systems of particles. I. The Green function and II. Scattering theory. Mat. Fys. Skr. Dan. Vid. Selsk. 2, No. 8 and No. 10 (1964-1965).

[37] Zislin, G.M.: The characteristic of a spectrum of the Schrödinger operator for molecular systems. Dokl. Akad. Nauk SSSR 128, 231-234 (1959)(russian).

[38] - : Discussion of the Schrödinger operator spectrum. Trud. Mosk. Math. Obšč. 9, 82-120 (1960)(russian).

[39] - : On the spectrum of the energy operator for a system of molecular type, on the space of functions having a given symmetry. Dokl. Akad. Nauk SSSR 175, 521-524 (1967)(russian); Soviet Math. Dokl. 8, 878-882 (1967)(english translation).

[40] - : Investigation of the spectra of differential operators of quantum mechanical many particle systems in spaces of functions of a given symmetry. Izv. Akad. Nauk SSSR, ser. mat. 33, 590-649 (1969)(russian); Math. USSR, Izvestija 3, 559-616 (1969/70)(english translation).

[41] Zislin, G.M. and A.G. Sigalov: On the spectrum of the energy operator for atoms with fixed nucleus on subspaces corresponding to irreducible representations of a permutation group. Izv. Akad. Nauk SSSR, ser. mat. 29, 835-860 (1965) (russian); Transl. A.M.S. Ser. 2, 91, 263-296 (english translation).

[42] - : On some mathematical problems in the theory of atomic spectra. Izv. Akad. Nauk SSSR, ser. mat. 29, 1261-1272 (1965)(russian); Transl. A.M.S. Ser. 2, 91, 297-310.